The Grafter's Handbook

R. J. GARNER

East Malling Research Station

THE GRAFTER'S HANDBOOK

New York
OXFORD UNIVERSITY PRESS

To Arthur William Witt
1864–1956
*for many years Chief Propagator
at East Malling*

Contents

9

CONTENTS

CONTENTS

CONTENTS

Illustrations

PLATES

FIGURES

Forewords

Foreword to the First Edition
In producing *The Grafter's Handbook*, Mr. R. J. Garner has had no easy task in steering a straight course between the fascinating myths of antiquity and the ingenious fads of the graft maniacs. He is, therefore, to be congratulated upon his clear exposition of this subject, which the fine line-drawings help to make abundantly plain. Both the text and the drawings reveal an enthusiast—but more than an enthusiast—addressing himself to fellow propagators. Whilst Mr. Garner's long association with investigations into the propagation of fruit plants naturally leads him to draw most of his examples from his experiences in this field, he reveals himself as a close observer and keen manipulator of many other plants—the ivy on the wall, the forced rose bush, the rhododendron, to mention but a few. The author has unconsciously succeeded in portraying himself wielding the tools, materials and methods which he describes so vividly. But here also is the craftsman weighing up scientifically the advantages and disadvantages of every detail of what he depicts. Only rarely is he tempted to record a process or a refinement just because it is delightfully complex or peculiar. Mr. Garner is at pains to bring his critical faculties into play and to show that this or that method of manipulation is the best suited to the particular plant, its structure, age, and condition.

The bibliography of over a hundred references testifies to the need for drawing together the somewhat isolated pieces of experimental information which have accumulated over the past thirty years. It was high time that these scientific findings should be correlated with the best traditional practices and summarized in an up-to-date *Grafter's Handbook*. The professional horticulturist, the amateur and the student will all find guidance and interest in these pages, since our knowledge of the purposes and effects of grafting have greatly increased its use and adaptation.

October, 1946 RONALD G. HATTON

Foreword to the Second Edition

The demand for a second edition of *The Grafter's Handbook* bears witness to its value both to the student in the field of propagation and to the skilled technician. During the past ten years the author has continued to pursue his own critical investigations, to collaborate with the physiologists, to keep abreast of the published work of others and to maintain his freshness as a teacher.

The valuable additions and adjustments in this second edition establish his right to expect that it will take its permanent place as the recognized textbook on the subject he so obviously enjoys and makes others enjoy. Here is a fine way of improving our techniques and inquiring into the principles upon which they are founded.

November, 1957 RONALD G. HATTON

Foreword to the Fourth Edition

This fourth edition is timely, not merely because previous ones have appeared at ten-year intervals but because the accumulation of new knowledge and the continued demand for the book requires it. *The Grafter's Handbook* is now firmly established internationally as THE encyclopedia of plant propagation by grafting; the Russian translation is widely used where English is less well understood.

With his characteristic thoroughness Mr. Garner has revised the text sentence by sentence, amending or supplementing each paragraph when new information seemed relevant and worth including. The familiar features, such as the clear drawings, are retained and the prose remains the essence of clarity. New references have replaced outdated ones, slightly shortening the bibliography.

23

Addicts of *The Grafter's Handbook* can be assured that it has been improved without losing the enthusiasm with which it was originally imbued. The book is of immense value to both the practical knifesman in the nursery and the professor of horticulture—what more can one say?

September, 1977 A. F. POSNETTE

Author's Notes

Note on the First Edition

I have written this handbook mainly for the practical horticulturist, but as it seems likely that students will also make use of it I have given numbered references to original publications throughout the text.

Having worked at the East Malling Research Station for twenty years, I have naturally drawn upon the information gained from the many experiments conducted by my colleagues and myself. I wish to acknowledge gratefully the encouragement I have received from my Director, Dr. R. G. Hatton, who has also contributed the Foreword, and the valuable assistance of my colleagues and other friends, far too numerous to mention individually. My thanks are especially due to Mr. David Akenhead, Director of the Imperial Bureau of Horticulture and Plantation Crops, for helpful criticism and for undertaking the onerous task of reading the proofs.

All the drawings are my own. For permission to incorporate those that have already appeared elsewhere and for most of the photographs, I am indebted to the Director of the East Malling Research Station and the Director of the Imperial Bureau of Horticulture and Plantation Crops. I have to acknowledge the permission of the Ministry of Agriculture to incorporate portions of text which I originally wrote for one of its bulletins.

In a concise handbook much must be left unsaid, and I realize that I am guilty of at least a few important omissions in my efforts to condense the relevant material. I should be most grateful for details of these omissions and for any other criticisms, so that in the future I can make this handbook more useful to my readers.

East Malling, 1947   ROBERT J. GARNER

Note on the Second Edition

In the ten years which have elapsed since the issue of the first edition many new methods and devices have come to light. An account of these I have incorporated in this new edition for the information of the practical man and the research worker.

Again I would thank my patient colleagues at the East Malling Research Station and the Commonwealth Bureau for counsel, criticism and permission to use photographs of technical processes. Readers of the first edition have kindly suggested ways of improving the handbook and I am most grateful to them. May I ask a similar favour from readers of the new edition? I would particularly thank Sir Ronald Hatton for again embellishing the work with a graceful foreword.

East Malling, 1958   ROBERT J. GARNER

Note on the Third Edition

The ancient art and modern science of grafting has continued to meet the call for greater simplicity in plant production and biologists' need for more highly skilled techniques for the unravelling of problems of growth and disease.

Simplicity of culture and accurate control of growth are now seen to be vitally necessary throughout the nursery life of the plant, in which the grafting process is but an incident. I hope that the additional information on 'replant' disease, and the nursery use of weed control, will prove useful. Methods of rootstock propagation by hardwood cuttings continue to advance and an account of recent progress will be found here.

I thank my colleagues at East Malling Research Station, and others, for helpful suggestions which have enabled me to clarify some of the expressions in the earlier editions.

East Malling, 1967   ROBERT J. GARNER

Note on the Fourth Edition

Thirty years ago when I wrote the note to the first edition I expressed thanks to colleagues and friends for their invaluable help. Now, in thanking them once again I wish also to thank my many new friends, in particular those associated with East Malling Research Station; the Commonwealth Bureau of Horticulture; the Hadlow College; the members of the International Plant Propagators' Society; and my fellow students and plantsmen throughout the world.

In half a century at East Malling I have witnessed a stupendous growth in the sponsoring of research and development. R. and D. grows by what it feeds on; the full diet is made available only by the efforts of gatherers and classifiers of information, by abstractors, editors and publishers. Aid for research must always include support for the recovery and dissemination of information.

Recent advances in instrumentation, photography and recording have helped to clarify our vision and widen our comprehension of complex biological factors and we can look forward to a resurgence of interest and an increased exploitation of grafting techniques, not only in horticulture but also in many branches of biology.

It is a special pleasure to thank Professor A. F. Posnette for his gracious foreword to this new edition of my book.

East Malling, 1978   ROBERT J. GARNER

Glossary

Adventitious: said of buds, shoots or roots arising out of order; not initiated by apical meristems; not in acropetal succession.

Bark: the outer layers of the rind, consisting largely of cork, serving to protect the inner rind and cambium (*see* rind).

Blanching: the exclusion of light from aerial parts of plants so that they become white and more tender (*see also* etiolated).

Bleeding: exudation from wounds, less likely while plants are in full leaf or dormant in late autumn or midwinter, or when dry at their roots.

Bud, to: to insert an eye, or bud, when bud-grafting.

Bud Grafting: grafting with a single eye or bud, detached from a shoot along with a portion of rind and, in some cases, a small slice of the wood (*syn.* budding).

Burr-knot: rounded swelling on a shoot, formed of incipient roots which may grow out in moist conditions. Sometimes confused with sphaeroblast.

Callus: healing tissue arising from the cambium at wounds.

Cambium: the layer of meristematic tissue between the wood and the rind from which further elements of both develop.

Chimaera: plant formed by coalescence of two or more forms, spontaneously or, rarely, as a result of grafting. Periclinal c., outer layer different from inner. Sectorial c., a sector different from the rest.

Chupon: orthotropic upright stem or vigorous erect shoot. In cocoa (*Theobroma cacao*) and some other plants the terminal bud breaks into three or five plagiotropic lateral fan-branches; the trunk produces only chupons which form further series of fans. Chupon scions continue to behave as chupons.

Cion: old spelling of scion; sometimes used in U.S.A.

Clone: vegetative progeny of one plant.

Compatible: said of plants which, when grafted together, form a good, sound and permanent union.

Cork Cambium: (*see* phellogen).

Cultivar: internationally-agreed technical term for what, in English, is known as a variety.

Disbud, to: to remove unwanted buds or young shoots.

Double-worked: twice-grafted, thus composed of three parts, rootstock, intermediate and upper scion (*cf.* single-worked).

Epicormic Shoots: shoots from buds dormant more than one season. Term normally reserved for shoots developed from buds long dormant. Not to be confused with adventitious shoots.

Etiolated: said of parts of shoots grown in complete darkness, as in the etiolation method of layering. Such parts are white or pale yellow and develop no leaves within the region of etiolation. Distinct from blanching (*which see*).

Eye: a single bud or group of buds.

Fan-shoots: plagiotropic horizontal shoots. Growths from fan-scions normally remain horizontal but may produce chupons following severe pruning.

Frameworking: methods of reworking established trees involving retention of existing framework of branches (*see also* topworking).

Free Stock: rootstock raised from seed.

Graft: where the scion meets the stock; the completed operation of grafting; the union; a term often wrongly used for scion.

Graft, to: to prepare and place together plant parts so that they may grow together.

Incompatible: said of plants which, when grafted together, fail to form a lasting union.

Intermediate: a scion interposed between a stock and a second, so-called upper, scion. A scion grafted at both ends either simultaneously or successively (*see* double-worked).

30

Juvenile Phase: initial period when plant cannot be induced to flower, prolonged in some conifers. Symptoms sometimes regained by very fast growths.

Labile Phase: growth phase between juvenile and adult; unstable; readily undergoing change.

Liane: climbing or twining plant, particularly in tropical forest.

Lignification: the process by which soft (herbaceous) stems become hard and woody.

Line-out, to: to set young rooted plants in rows in the open nursery to grow larger, or for working.

Mentor: a long-established adult scion used to influence a young seedling (Mitchurin, *Selected Works*, Moscow, 1949).

Meristem: tissue capable of growth, either primary from which new organs develop (primordia of leaves, stems, roots) or secondary by which special tissues are extended (cambium, phellogen).

Meristematic: pertaining to meristems.

Mother Tree: a tree selected as a source of scions, cuttings or seed.

One-year Bedded: plants which have been lined-out thickly for one year following propagation.

Paradise Stock: dwarf-growing vegetatively-propagated fruitful apple emanating from the Garden of Eden, used as a dwarfing rootstock. Now known that vigour range is not linked with mode of propagation.

Petiole: leaf-stalk.

Phellogen: meristematic cork-producing tissue in the outer region of the rind.

Polarity: tendency to develop from the poles, roots downwards stems upwards, making it essential to set 'the right way up'.

Raffia: the native Malagasy name for the fibre-like material obtained from the leaves of *Raphia pedunculata* and *R. vinifera*, used for tying grafts and for many other horticultural purposes.

Raphia: (*see* raffia).

Rind: all the tissues external to the woody core of stems and roots, often termed bark (*which see*).

31

Rootstock: root-bearing plant on which the scion is grafted (*see* stock).

Scion: part of a plant used for grafting upon the stock plant (*but see* double-worked).

Scion-rooting: development of roots from the scion so that the grafted plant has roots directly from both rootstock and scion.

Scion-wood: shoots from which scions are cut.

Single-worked: once-grafted, thus composed of two parts, stock and scion (*cf.* double-worked).

Snag: part of the rootstock left (temporarily) above the graft, usually used as support or protection for growth from the scion.

Snag, to: to remove the snag when no longer required.

Sphaeroblast: rounded swelling on shoot arising in the cortex, able to form adventitious buds. Often confused with burr-knot.

Staddle: a foundation of trunk and main branches (of rootstock or stembuilder) for grafting.

Stem Builder: a variety used as an intermediate stem piece to provide a good trunk for standard trees or to introduce resistance to disease or winter injury.

Stock: the plant, or plant part, receiving the ingrafted scion (*see* rootstock).

Tissue: the substance, structure or texture of which the plant body, or any organ of it, is composed.

Topworking: reworking an established tree after removing the head of the tree back to the bases of the main limbs.

Uncongeniality: exhibited by plants grafted together which are not completely compatible, sometimes shown by large swellings or overgrowths at the graft union. Severely dwarfed growth accompanied by symptoms of ill-health traceable to grafting.

Understock: a term sometimes used instead of 'rootstock', especially descriptive of rootstock in a double-worked plant.

Union: (*see* graft).

Wood: water-conducting and strengthening tissues forming the skeleton of the stem or root. Secondary thickening of these tissues results in the condition of 'woodiness' (*see*

lignification).
Work, to: (*see* graft, to).
Worked: grafted.
Xylem: botanical name for wood.

Reasons for Grafting

Plants may be grafted in a multitude of ways and for many different reasons. The art may be exercised merely as a pastime, but grafting is usually employed to gain one or more of the following objectives:

1. To propagate, or to assist in propagating, plant varieties not otherwise conveniently propagated.
2. To substitute one part of a plant for another.
3. To join plants each selected for special properties, e.g. disease resistance or adaptability to special conditions of soil or climate.
4. To repair damage, to overcome stock/scion incompatibility, and to invigorate weakly plants.
5. To enable one root system to support more than a single variety or one branch system to derive from more than one root system.
6. To elucidate problems of structure, growth, and disease.

The Plant as seen by the Grafter

The more the grafter learns about plant structure and behaviour, the greater become his chances of success. To increase in knowledge in these matters the novice must study plants closely, give heed to qualified teachers and refer discreetly to a few well-chosen text-books. Even for those who remain satisfied merely to employ the art to achieve an end, it is still essential to know something of plant construction.

The drawing opposite serves to show any tree, or woody plant, that may be grafted. At the lower end there are small roots and at the upper there are twigs which bear leaves in due season. The small roots concentrate into larger roots, these into the stem or trunk, the trunk divides into limbs or branches which separate into smaller and smaller branches, which finally end in buds. The grafter may regard these structures as pipes of varying thickness, where the lagging on the outside of the pipe is the rind of the root or stem, the content of the pipe is the hard wood or skeleton, and the metal of the pipe itself is the cambium. The dotted lines in the drawing represent the cambium.

The rind thickens with age, and is thicker in roots than in stems of equal age and diameter. These facts are emphasized in the illustration where sections of root and stem are seen as through a magnifying-glass. When the tree is in active growth the rind can be lifted from the wood, thus clearly revealing the position of the cambium. At other times the whereabouts of the cambium must be discovered by close inspection of the section and, to ensure good cambial contact, the thickness of the rind of each component must be kept in mind when grafting.

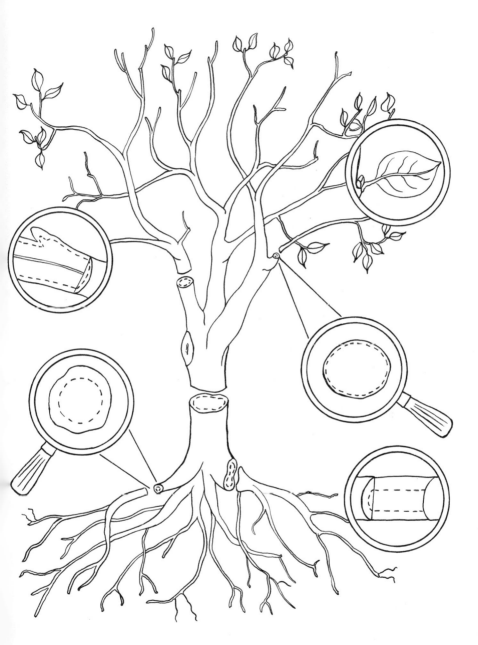

The Plant as seen by the Grafter

CHAPTER I

Grafting in Nature and Antiquity

Natural grafts are sufficiently common to be readily accessible for observation by a student of grafting. By studying grafting in nature one can learn the essential characteristics which are necessary for success and the almost unlimited technical possibilities. Some species are particularly prone to natural grafting and, among the commoner trees, beech (*Fagus sylvatica*), whether growing naturally or in hedge form, provides numerous examples (Plate 1). The elm (*Ulmus campestris*), ash (*Fraxinus excelsior*), common maple (*Acer campestre*), the poplars (*Populus* spp.), conifers including cedar (*Cedrus atlantica*) and pine (*Pinus sylvestris*), shrubs, and climbers have all been reported as forming unions between their own branches by natural grafting. Scots pine (*Pinus sylvestris*), which is not easily grafted artificially, owing to a profuse exudation of resin from wounds, nevertheless forms efficient natural grafts between its branches (Plate 12). Similarly some latex-yielding plants, notably *Ficus* spp., readily form natural grafts (Plate 2). Probably the most readily available source of specimens is the common climbing ivy (*Hedera helix*) (Plate 2); moreover, these examples can usually be collected for dissection without involving the destruction of the plant, merely because the presence of the many efficient unions enables parts which have lost direct connection with the root to be sustained via the junctions brought about by grafting.

A careful study[92]* of natural grafting in climbing ivy shows that the sap can pass from one stem to another across the union and then up and down in the second stem. This

*Numbers in text relate to references on pages 292–300.

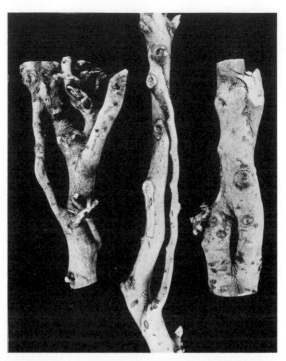

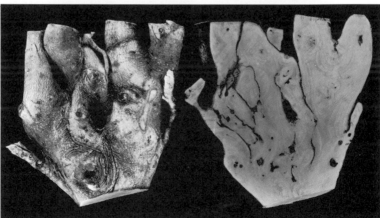

1. Natural grafts from a beech hedge

Dark spots in the section indicate inclusions of dead rind commonly seen in natural grafts.

explains why the severing of ivy stems has merely a local effect upon growth and why it is necessary to break all living connection with the soil, if the plant is to be killed. Upon close inspection it will be seen that where stems have come in close contact, and especially where they have crossed, they are joined together, so that in established ivy there is a wonderful network of stems and the upper branches are able to draw nutrient solutions from far distant parts of the plant. These unions are made when the stems are young, soon after they first come together, for it is seen that older stems, loosened from their support and placed in contact, form no union. It appears that natural grafting depends first upon contact whilst the bark is still thin and then upon firm anchorage. These conditions are fulfilled in climbing ivy by contact of the stems when first put forth and their permanent anchorage against their fellows. Though a union may not take place immediately, the bark is unable to thicken at the point of contact and a union is perfected over a period of years, so that established unions and mature stems grow old together.

Another climber, with well-anchored stems but which has been examined in vain for natural grafts, is the climbing hydrangea (*H. petiolaris*). With this subject the failure to form graft unions when in close contact appears to be due to the peculiar nature of the bark, which is fibrous and readily peels in flaky strands. The internal living layers of such stems are kept apart by comparatively thick layers of loosely-constructed bark and are not affected, apparently, by pressure of neighbouring stems.

In climbing ivy the process of fusion, or union, begins with compression and it seems that this constant and increasing pressure, combined possibly with slight but powerful movements occasioned by adjustment to meet strains, ruptures the outer bark. This is thought to release tension in deeper layers and to initiate growth of tissues in rind and cambium. Now this is exactly the procedure in wound healing. The restricting envelope being released by the wounding, the lower living layers—capable of growth—bulge out and eventually meet and join to form a complete covering. In the ivy stem the rupturing of the bark is followed by growth from the inner rind, the phellogen, which is usually immediately followed by growth of the cambium in the form of protuberances. This happens in both stems, and the protuberances push aside the disintegrating outer bark, as shown diagrammatically in Fig. I, and meet each other to form a complete layer of cambium and new rind and bark around both stems. Put briefly, grafting, whether natural or artificial, is the healing in common of wounds, but successful grafting depends also, as will be seen in Chapter II, upon the compatibility of the plants joined together. Though unusual in nature, where opportunities of close proximity under conditions suitable for natural grafting rarely occur, the joining of differing species is commonly engineered by man in horticultural practice. Experiments have proved that the angle at which the stems cross each other has a marked effect upon the amount of solutions which may be passed through the union[92]. This effect is less noticeable in long-established unions, but is an important point in considering grafting technique.

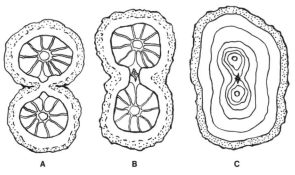

Fig. I Natural grafting between two young stems

(diagrammatic transverse section)

A. The outer rind is ruptured and growth in the cambial region opposite the wound is accelerated. B. The protuberances of new tissue meet and the cambia join. C. The union is complete and the two stems appear as one.

Natural grafting in non-climbing plants, such as timber trees, appears to be more common in those which have smooth or thin bark, at least in the early stages of growth. Opportunities for grafting occur where branch systems are complicated or congested or where boughs droop and rest one upon another. Mere rubbing together and wounding by continuous friction will not bring about union and it is clear that firm anchorage of the parts is vital. Presumably this is the reason for the comparatively large proportion of the grafts occurring where one bough becomes wedged in the fork formed by a bifurcation of another. A form of natural branch grafting is encouraged by fruit growers to provide support for limbs which may later carry heavy crops. This operation is performed[125] by twisting together two fairly strong young lateral shoots from the limbs which are to be braced. It is often necessary to brace the limbs temporarily with rope or stout cord and the entwined shoots are firmly tied with string to keep them in close contact (Fig. 2). In a year or two a strong natural and permanent brace is formed.

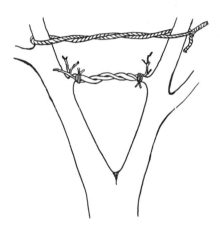

Fig. 2 Bracing limbs by encouraging natural grafting
The limbs are held by a stout cord until the two young shoots have united by natural grafting.

In formal gardens of the eighteenth century natural grafting was brought about when trees were twisted and tied together in imitation of columns and balustrades[119].

So far the above-ground parts have been considered because they are the most readily observed, but grafting underground between roots is even more general[54], though less frequently seen because of limited opportunities. Opportunities do occur, however, when trees are uprooted by gales, orchards are grubbed or soil erosion occurs on hillsides, river banks or road cuttings. In forests it has been observed[13] that trees are able to form highly efficient grafts between roots of the same species so that a cluster of trees may be so thoroughly joined that all but one tree being cut through at ground level, the roots of the whole group will survive. When fruit trees are grubbed, often because of overcrowding, root grafts are frequently seen and, where one system is intimately mixed with another, intergrafting is common. Thus trees having roots of different varieties may become amalgamated, if only in small degree. The frequency of natural grafting between roots is not surprising; the conditions found to be necessary for success in stem grafting are here to hand, namely perfect anchorage and little protective cork in the rind.

Roots may join at any angle but are seen to be most efficient, judged by thickening, where they approach each other at an acute angle, as in stems (Plate 2), but grafting also occurs between roots meeting at right angles and even where they approach from opposite directions. In these last it has been noted that the tissues, appearing in the new annual ring, turn and twist in a wonderful manner so that they are ultimately directed in the same way.

Natural grafting between roots of neighbouring trees may be a means of spreading disease in a forest or plantation. 'Oak Wilt' (*Endoconidiophora fagacearum* Bretz.) spreads from tree to tree, of the same species, through grafted roots[78] and the only practical means of control is to kill a margin of trees around the infected area. Observations in Californian citrus orchards[8] revealed that the slow spread of certain virus diseases was chiefly due to root grafting between infected and healthy trees. In the process of virus indexing by sensitive indicators valuable healthy clones may be inadvertently discarded because the indicator has become infected by root-graft connection to a nearby infected plant. Wider spacing of the plants reduces this risk[55, 74].

Root grafting is very common in the rubber plantations of Malaya. In 1950, some five years after trees had been cut down by the invading army to within less than a metre (3 ft.) above the ground, the leafless stubs were opened up for tapping. Yields were comparable with those of neighbouring

2. Natural grafting between stems and branches

*A. Ivy (*Hedera helix*) stems severed three years earlier draw their sustenance via the numerous junctions. B. Creeping Fig (*Ficus pumila*) entirely engulfing its support. C. Cedar (*C. atlantica*) supported by a strong natural bridge. D. Closer view, showing increased development above the graft and thickening towards upper end of the bridge.*

undamaged trees to which they were joined below ground (Plate 3). When thinning plantations it has been customary to 'frill girdle' and poison intermediate trees with sodium arsenite or the like. Where there are root grafts the poison may pass to neighbouring trees and the whole plantation may be severely crippled, if not entirely destroyed.

When completely girdled trees continue to grow and develop over many years this is due to establishment of root grafts with non-girdled neighbouring trees. Authenticated examples have been quoted for Douglas Fir (*Pseudotsuga menziesii*), Oregon Oak (*Quercus garryana*), and Pacific Madrona (*Arbutus menziesii*)[115].

Trees growing in loose sandy or peat soils are poorly anchored and in exposed situations are subject to 'wind-throw', but cases are reported[54] in which a firm stand was attributable to vigorous and widespread establishment of root grafts giving mutual support.

It is perhaps of interest to note that natural root grafting has occurred between clove trees in Zanzibar[110]. The clove is extremely difficult to graft artificially and that seen in nature was between young seedlings, or between a seedling and an older plant. This suggests that seedlings in the juvenile phase may be more amenable to grafting than more mature plants and supports the view that young seedling material grafts more readily than adult (*see also* page 54).

It seems reasonable to suppose that artificial grafting followed upon observation of natural grafting and doubtless took similar form to those methods now grouped under the heading 'Approach Grafting', described in Chapter VI. Grafting with detached scions must surely have been a later invention, probably encouraged by the observation of damaged approach grafts. Grafting with detached scions has been practised for thousands of years; it was used by the Chinese before 2000 B.C., and even earlier by the first known citizens of Mesopotamia. In ancient Greece the technique was almost commonplace. Many weird ideas were current then, as now, and some proposals had an intriguing, magical quality. The seedling plants found flourishing in the crotches of ancient trees were described as grafted, and very strange suggestions were made for the improvement of trees by uniting their

Courtesy of E.D.C. Baptist and Director, R.R.I., Malaya

3. Natural root grafts provide mutual support

A. Rubber trees joined by natural root grafts.   B. Closer view of A.   C. Although the tree on the left was beheaded five years earlier it continues to yield latex via a root graft connection with the intact tree on the right.

qualities. Observation and experiment have long since proved many of these early suggestions incapable of realization but, nevertheless, they make interesting reading. It is clear that the ancients had learned the immense possibilities of improving plants by grafting and the modern grafter would do well to become imbued with some of their inquisitiveness and enthusiasm. That their ideas exceeded the bounds imposed by laws of compatibility is now recognized. It should be remembered that those most skilled in the art were not the most likely to record their ideas in writing. Scholarship was not commonly an attribute of husbandmen.

In spite of the Jewish rabbinical law forbidding hybridization, and the orally-transmitted law prohibiting grafting in order not to raise hybrids, by the time St. Paul was writing his epistles* not only was grafting well known in commerce but the possibility of reaction between scion and rootstock was also recognized. Had we now access to horticultural literature of the period, we should probably find that innumerable methods of grafting were known and used. The sixteenth and seventeenth centuries in England were distinguished by much sound horticultural writing including detailed and reliable information on grafting[85]. For example one excellent little book[25] tells us to bud our oranges with an inverted T incision to obtain good success, discourses upon leaving the wood in the bud shield and upon grafting in spring with properly dormant scions. The mode of joining was, for all practical purposes, well understood. In the words of that time: 'The sap, arriving at the outside of the wood, there to generate a new orbe (annual ring), doth in the restagnation seize upon and conjoine itself to the moist part of the graft (scion), that is affinity unto it.'

* Romans XI, 16–24.

CHAPTER II

Compatibility and Cambial Contact

Success in forming a permanent graft union between plants or plant parts depends upon two things. One of these is intimate relationship or affinity, often termed compatibility, between the subjects joined together, and the other, contact between their cambia or other meristematic tissues. The characteristic continuous cambium and related elements in the dicotyledons lend themselves to the grafting process. On the other hand, the scattered arrangement of the vascular bundles in the monocotyledons and the absence of secondary thickening involve difficulties for the grafter. These, however, have not always deterred investigators. Virus diseases have been transmitted from one liliaceous plant to another across a graft union. There have been reports from Russia[102] on the grafting of cereals in hybridization work and some success has also been achieved elsewhere[96,56] with grasses, sugar cane and bamboos. Nevertheless, general grafting practice is concerned with the dicotyledons and the conifers, both of which possess a true cambium layer in their stems and roots. Some dicotyledonous lianes, however, possess a stem structure in part resembling that of monocotyledons[89]. The vascular bundles in young stems of *Piper nigrum* L., for example, are widely spaced and there is no continuous cambium. As the stem ages, an interfascicular cambium arises and, for a while, becomes continuous with the cambium of the outer vascular strands. At this stage, secondary thickening takes place and stems may be grafted successfully. Later in the development of the tissues, the interfascicular zones become lignified and take the form of fibre flanges disrupting the continuity of the cambium. It is the development of these flanges which is responsible for the poor survival of grafts that were initially successful[46].

49

COMPATIBILITY

The importance of close relationship for success in grafting is well known, and no one would seriously set out to graft a member of one botanical family with a member of another. Cases have been reported of successful grafts between members of different families, but such isolated successes have frequently been between annuals or biennials. Between unrelated woody perennials it is doubtful if a definitely compatible union has ever been established.

The botanist's classification serves only as a rough guide to compatibility, for it is founded upon the reproductive characters, and experienced grafters have learnt that this is not a reliable guide. Something more than kinship is required for a good union[17]. On the other hand, the achievement of graft unions between distantly-related plants does not seem so strange when it is remembered that successful grafting depends upon similarities between vegetative characters and that these are not necessarily correlated with the reproductive characters. Permanent unions are not uncommon between one genus and another; pear (*Pyrus communis*) will form a lasting union with hawthorn (*Crataegus oxyacantha*) and with the medlar (*Mespilus germanica*) but the joining of species within a genus is more common, though many of such combinations are short-lived.

Whilst the botanist's classification is not a completely adequate guide for the grafter, there are indications of a positive correlation between interspecific hybridization and grafting success. Working with nine species of cacao (*Theobroma*)[1] it was found that species which produce hybrid progenies, or at least hybrid seeds, can be reciprocally grafted, whereas interspecific graftings which fail also fail to hybridize. Whether hybridization and grafting have a common physiological basis remains doubtful but work with the genus *Trifolium* [30] also indicates that a correlation exists. Intergeneric apple (*Malus*) and pear (*Pyrus*) crosses[128] show that the pear 'Fertility' is more successful than many others. Studies at East Malling[51] using eighteen apple rootstock clones and three pear cultivars also show that 'Fertility' pear has an exceptionally high graft compatibility with apple. In turn, one of the eighteen apple clones (M.16) was itself exceptionally compatible with each of the three pears.

There are many examples of seedlings of a single species, closely resembling one another in externals, behaving very differently when grafted with one variety. When individual seedlings are used, the phenomenon is obscured because the occasional failures are attributed to faulty manipulation or to a bad season. Only when each of these individual seedlings is kept apart, propagated vegetatively, and grafted in quantity does the behaviour become strikingly clear. For example, four plum seedlings selected from one batch of seed of a single species and each multiplied vegetatively, and all budded with one variety of peach, behaved very differently. Selections 1 and 2 formed perfectly compatible unions, though differing in vigour. Selection 3 made vigorous young trees, all of which had died by the end of the fifth year from budding, whilst trees on selection 4 failed to survive beyond the maiden year. This, and many similar experiences, have taught that each combination must be tested by grafting before its compatibility can be determined*. Present investigations also suggest that detailed microscopic examination of the internal structure of the components of a proposed union may at some future date afford a reliable technique for forecasting compatibility between individual plants. Though it has not been possible to predict incompatibility in advance of grafting, examination of the structure of the union of one-year-old trees[64] under the microscope has revealed abnormalities well in advance of externally visible effects.

A mass of literature on stock/scion incompatibility or uncongeniality has accumulated in recent years. This has been reviewed in considerable detail[2,94] and set in order. Incompatibility can take many forms from slight symptoms of ill-health to complete incompatibility, where no union is formed when the components are budded or grafted together. It is clear that factors other than incompatibility may produce very similar effects and it is therefore necessary to define what

* See Appendix II for table of reliable combinations with commercial plum rootstocks.

is meant by the term when applied to the graft union.

The term 'incompatible' should be reserved for distinct failure to unite in a mechanically-strong union, failure to grow in a healthy manner, or premature death, where such failure can be attributed with a reasonable degree of certainty to differences between stock and scion. In a compatible combination the actual junction is at least as strong mechanically per unit of cross section, as revealed by breakage tests, as the parts immediately above or below the union. Where the wood of one of the components is brittle and may be broken cleanly, rather like the root of a carrot, any breakage close to the union may appear to the casual observer as though the severance were actually at the union and the combination may be regarded, mistakenly, as incompatible. An example of this is seen in the apple rootstock Jaune de Metz (Malling 9) which is readily broken with a clean break somewhat like the fracture at the junction of varieties mutually incompatible. Inefficiently staked trees of the common apple varieties on this rootstock are liable to break close to the union during high winds, but the many casualties examined have shown that this has occurred below and never at the union. If the part above the fracture is sawn vertically, as in Fig. 3, and the

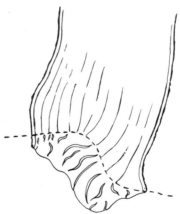

Fig. 3 Smooth break below the union

Longitudinal section of a tree of Cox's Orange Pippin apple worked on M.9 (Jaune de Metz) which has broken cleanly below the union, in the rootstock tissue, simulating severance at the union due to incompatibility of the breaking type.

4. Swellings at the union

Two fifty-five-year-old cherry trees (P. avium) grafted at 1·8 m (6 ft.) in the nursery. The variety 'Old Black Heart' has produced a large swelling above the union, typical of this variety; the other, 'Kent Bigarreau', has a smooth union. Both trees have normal health and continue to bear satisfactory crops.

sections left in a dry place, the rootstock, containing less lignified tissue, will shrink and crack before the scion-piece, and the exact position of the union is then readily discerned.

The term 'incompatible' should not be applied to cases in which treatment and environment would appear to be the most likely cause of failure, or to cases where abnormalities, e.g. union overgrowths as seen in Plate 4, are unaccompanied by failure to survive.

The most reliable indication of incompatibility is breakage at the point of union, particularly when the combination has survived beyond one growing season and where the break is complete, smooth and unsplintered (Plate 5). It should be noted that a clean break can also occur at the junction of components, normally compatible, owing to the insertion of a bud or graft into a stock having a very thick rind. In such cases[130] it appears that the rind restricts expansion of the base of the maiden shoot, and junction between the scion-shoot above the constriction and the stock is prevented by the unyielding layer of cork in the rind of the stock. Losses due to

Copyright East Malling Res. Stat.

5. Breakage at the graft junction

A well-established and vigorous plum tree which has broken at the graft, leaving a smooth, non-splintered surface, a reliable indication of stock/scion incompatibility.

this may be prevented by paring the rind of the stock as described on pages 150, 162 and 283.

COMPATIBILITY IN RELATION TO AGE

There is no indication that a plant in the mature growth phase changes in degree of compatibility with any given scion as it ages. Where this appears to have happened it may well be due to the development of systemic disease or nutrient deficiency. There is, however, some evidence that plants may be more readily grafted in the seedling and juvenile stages than later. In adverse conditions in the nursery, seedling rootstocks are generally found to provide better stands of young trees than do selected clones vegetatively propagated. Further, plants which are extremely difficult to graft when mature have been readily united in the very early seedling or cotyledon stage[80,135]. This greater grafting success with seedlings may be due in part to growth speed rather than to growth phase alone. A similar vigour effect is seen in the regenerative capacity of cuttings, whether such vigour is associated with the juvenile phase, adventitious origin or invigoration by rootstock, pruning or environmental control[50]. It is suggested that high regenerative vigour, with postponed flowering, will contribute to both cutting regeneration and grafting success.

DELAYED SYMPTOMS OF INCOMPATIBILITY

Symptoms of incompatibility may be delayed for many years. This is seen in pear plantations where certain varieties exhibit 'delayed incompatibility' with quince[39]. Sometimes these combinations grow and crop normally for many years, but there is considerable variation in the trees of a single planting. This may extend from the normal tree down to the tree that breaks clean off at the union a few years after planting. It is common to find plantations of these varieties composed of trees which vary in age from twenty years and more to the newly-planted maiden. This condition is reached by annually replacing the casualties, and apparently this procedure may well continue as long as the same combination of scion and rootstock is employed. The moment of expression of incompatibility is largely determined by local conditions. The less favourable the conditions the greater the disharmony. In good nursery environments symptom expression may be delayed only to become clear after transference to the field. There is a correlation between incompatibility and susceptibility to frost damage in the scion[32,65].

The loss of trees by breakage at the union is due to a form of incompatibility in which the fibres of scion and rootstock fail to interlock[27]. A forty-year-old Colorado White Fir (*Abies concolor* Hildeb.) grafted on European White Fir (*A. alba* Mill.) snapped cleanly at the union. The tree was apparently healthy and had grown normally, with no interlocking cells between stock and scion. Contact between tracheids and sieve tubes was intimate but horizontal. These side-by-side cells had apparently conducted solutions quite efficiently. This incompatibility at the union can be overcome by the use of an intermediate stem piece or bridge of a variety compatible with both scion and rootstock. The process is known as double-working or bridging*.

* See Appendix I for list of pears improved by double-working.

Sometimes a grafted plant which first exhibits symptoms of incompatibility some years after grafting grows abnormally vigorously in its first year. This has occurred with lilac (*Syringa vulgaris*) on privet (*Ligustrum vulgare* or *L. ovalifolium*)[14], and with sweet and acid cherries (*Prunus avium* and *P. cerasus*) grafted on the mahaleb or St. Lucie cherry (*P. mahaleb*) and on the dwarf-growing *Prunus wadai*. Darwin[21], as early as 1875, records that 'varieties grafted on very distinct kinds, though they may take more readily and grow at first more vigorously than when grafted on closely-allied stocks, afterwards become unhealthy.' Exceptional scion vigour, preceding incompatibility symptoms, is also reported[106] for combinations of Broad Bean (*Vicia faba*) with Lupin (*Lupinus luteus*). These, and other observations suggest that exceptional growth, exceeding that of either component alone, may be a symptom of incompatibility.

POSSIBLE CAUSES OF INCOMPATIBILITY

The cause, or causes, of incompatibility may still be regarded as undetermined; the evidence is as yet largely inadequate and often conflicting. Some investigators[19] regard equal growth rates and cycles as essential accompaniments to a congenial graft union, based on the normal behaviour of the plants before grafting. However, there are many apparently congenial combinations which show growth differences between stock and scion just as great as others whose failure has been attributed to this variation in growth. Conversely, failures are seen to occur between plants of very similar growth, and therefore growth-rate differences alone cannot be regarded as the basic cause of incompatibility. Moreover, the time of the operation in relation to season neither delays nor hastens the onset of symptoms of incompatibility[47].

Attempts to associate incompatibility with physiological or biochemical differences between stock and scion have so far also given inconclusive results but work in Holland with Cucurbita grafts[22] has markedly advanced ideas in these fields. With *C. sativus* as the upper scion, *C. melo* the intermediate, and *C. ficifolia* the rootstock, the combination is incompatible. Yet with no *C. melo* in between the combination is perfectly compatible. The suggestion is that essential substances, of enzymatic or hormonal nature, made by the upper scion, are altered in their passage through the intermediate and so lead to the failure of the rootstock. This idea was tested by reducing the intermediate to a very thin transverse section of little more than 1 mm (1/25 in.) (Fig. 77). When this was done the hitherto incompatible combination was quite successful. Apparently the short distance of travel of the metabolites concerned did not allow time for a change in their nature.

It has been possible to postpone the onset of these ill-effects of incompatibility by the retention of some leaves on the rootstock and it is suggested that an active principle emanating from these leaves maintains health in the composite plant. Another suggestion is that rootstock leaves supply carbohydrates to the root system which might otherwise be starved by faulty union with the scion.

In contrast, a previously compatible union has become incompatible following the addition of scions of a third variety[95] and in this case the retention of leaves by the intermediate did not postpone union failure. The tissues of the rootstock below the union were the most seriously affected; there was a breakdown of phloem tissue and the roots contained little or no starch. In healthy compatible trees starch is well distributed through root and stem. The addition of another scion, containing a virus to which it is tolerant, to an existing compatible combination of virus-sensitive plants may lead to breakdown of tissue and failure of one or both of the graft unions.

Union failure between the Persian walnut (*Juglans regia*) and the black walnut (*J. nigra*) some fifteen to twenty years after grafting has been reported in America[120] and Europe[53]. The trees grow well at first but later cease shoot growth and drop their leaves abnormally early. The rind (bark) becomes discontinuous at the graft union and dead bark remains between stock and scion (Plate 6). The bark inclusion, termed 'black line', first appears when the trees change from the young vigorous to the full fruiting phase[91].

Discontinuity in the rind alone, with no involution of bark between the wood of stock and scion, indicating normal cambial function towards the xylem, does not markedly

6. Long-delayed incompatibility

*A. Vertical section of the union of a twenty-four-year-old walnut (*Juglans regia*) on* J. nigra. *Discontinuity first shows in the twentieth year. Arrows indicate the change. B. Vertical section of the union of a fourteen-year-old plum, Marjorie's Seedling on Myrobalan B. Discontinuity from the fifth year, resulting from the use of a virus contaminated scion in the nursery.*

impede overall growth nor lead to mechanical weakness at the union. It appears that the discontinuity is not absolute but is a result of 'make and break' permitting functional contact between stock and scion, revealing a tolerable degree of incompatibility (Plate 7).

Where the symptoms of incompatibility are delayed, as in certain pear/quince combinations already mentioned, the shoot growth of the tree may be at least as good as in a sound tree produced by double-working or in a tree of another variety completely compatible with the same quince, in fact often superior, so that there is in these cases no suggestion of interference with the flow of essential solutions or plant foods across the union, in spite of the presence here of horizontal plates of undifferentiated tissue. Thus such incompatibility is not comparable to a constricting girdle, or a blocking of the passages, for it seems that materials readily pass across the junctions, and it therefore appears that mechanical weakness, due to lack of interlocking fibres, is the chief or only limiting factor. On the contrary, in thoroughly compatible unions, where the fibres are firmly interlocked, there is no free passage for all materials or intermixing of tissues. Each component retains its own special characteristics right up to the junction, as can be clearly seen when plants with coloured cell sap are grafted upon those with colourless cell sap, the colour ceasing abruptly at the union. Further, where an intermediate stem piece, having coloured sap, is inserted between a colourless stock and upper scion, the colouring begins and ends abruptly with the intermediate (Plate 8). In grafting experiments this fact has been used to demonstrate the absence of intermingling or 'overflowing' of the tissues. When sectioned, such combinations reveal that the original parts retain their identity to within a layer of cells of the junction and though one part may grow faster than another, so causing a bulge, there is no overlaying of one component by another. When the components have distinctive barks the junction is exactly defined (Plate 8). Overlaying by the scion has been said to occur in bud-grafted rubber (*Hevea brasiliensis*)[24] but closer observation fails to substantiate this. Again, in the olive (*Olea europaea*), the character of scion-rooting may deceive the casual observer. A peculiar feature of

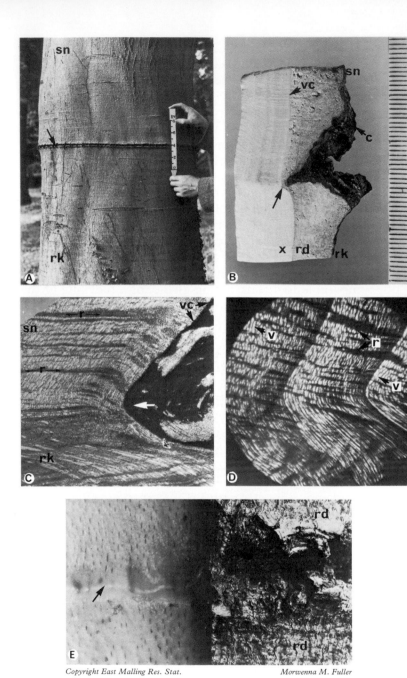

Morwenna M. Fuller

7. Tolerable incompatibility

7. Tolerable incompatibility

A. The trunk of a well-grown beech tree about 100 years old showing the position of the union (arrow) between the scion, Fagus sylvatica purpurea, *and the rootstock,* F. sylvatica. *B. A sample cut from the trunk of the tree shown in A. The darker grey wood of the scion denotes a pink coloration in the xylem tissues of* F. sylvatica purpurea. *Some degree of incompatibility is indicated by partial discontinuity in the rind. The living tissues of the rind, in the plane shown in this photograph, are united near to the vascular cambium (arrow). C. Radial longitudinal section across the union showing involution in the cambial zone (white arrow). D. Radial longitudinal section across the union of the xylem tissues cut in a different plane from that of the section shown in C. Vascular continuity exists between stock and scion but disorientation of the cells is apparent along the line of the union. E. A close-up view of the trunk showing the broken tissues of the rind. Removal of the rind from the zone on the left has revealed the surface of the wood. The union of the xylem tissues is clearly visible (arrow).*

Key: *sn = scion; rk = rootstock; x = xylem; rd = rind; c = cork; vc = vascular cambium; r = ray; v = vessel.*

Eleanor C. Thompson   Copyright East Malling Res. Stat.

8. Each component retains its identity

*The smooth-barked English Walnut (*Juglans regia*) patch budded on to the rough-barked Black Walnut (*J. nigra*). The greatly-enlarged 'patch' is still clearly visible and there is no intermingling or overflowing at the junction.*

olives is the replacement of the original roots by 'cords' of downward-growing aerial roots. The rootstock—on the rare occasions used—merely serves as a nurse until the scion has become connected with the soil, when it dies and rots away.

CAMBIAL CONTACT

Provided the plant parts placed together are living and mutually compatible, then the only remaining essential to success is that the cambia or other meristematic tissues be in contact at least in some degree, or so close together, that they achieve contact in conditions favourable to further growth. Obviously other factors may aid success, but these are not essential in all cases; at the same time compatibility and cambial contact alone do not guarantee success, because adverse conditions may bring about the death of one or other of the partners.

The position of the cambium within the plant has already been described. It has been shown that this cambium is present in the form of a continuous tube extending throughout the roots and stems and that this tube has wood within it, which is not able to grow, and a rind outside which conducts elaborated plant foods and other substances. This rind contains cork cambium (phellogen) which builds a continuous layer or skin of protective material commonly known as bark in woody plants. It is true that in young stems new growth, in the form of callus, may occur in the region of the pith—in fact this may be strong enough to force new grafts apart—but grafters are chiefly concerned only with the true cambium, which extends through the plant in a tube of varying size and shape. Now, any method of carpentry which brings the cambia of the components together, under conditions sufficiently favourable for growth, may achieve success. Though it appears that only a small area of contact is really necessary, a generous and firm contact usually results in an earlier establishment of an efficient union.

If the cambium is regarded as a tube or cylinder it is then clear that one of the simplest means of making contact between two tubes is to cut through both transversely and to place them end to end. Some system of splinting would be necessary to immobilize the parts until the formation of a strong union (*see* Abut graft, page 197). The simplest practical method of preventing independent movement of the parts is to prepare the components, not by cutting transversely, but by long slanting cuts so that they splint each other. This simple method of grafting is known as the splice (Fig. 56). Some operators consider that a simple splice tends to slide under pressure and have suggested the use of a notched splice rather like the arrangement used by carpenters in lengthening beams required to take thrusts. In grafts of this kind which are firmly tied, such modifications are usually unnecessary. The simple splice requires that the parts be held in position whilst tying. With one operator working alone this is no great inconvenience, but with two or more sharing the work it becomes impracticable. For this reason a modified splice graft known properly as the 'whip-and-tongue', or more shortly the 'whip', graft has been universally adopted (Fig. 57). The use of interlocking tongues enables the grafter to set the scion in a stable position and leave it there for the tyer to complete the work.

Whilst it is a comparatively simple matter to place the cambia in contact where the stems are of similar size, it requires rather more care to achieve cambial contact between stems of different size and age. In one-year-old stems the rind is thin and the cambium is close to the outside of the shoot, but in older stems the rind is much thicker and the cambium much further from the surface. Thus the matching of the outer edges of the rind will not bring about cambial contact and it is always necessary to place the inner edges of the rind together. This is shown diagrammatically in Fig. 4. The rind of roots is considerably thicker than that of stems of the same size and in stem/root grafting it is vitally necessary to place the inner edges of the two rinds in contact.

Any grafting process which involves lifting the rind from the wood when the cambium is in active growth will itself reveal the position of the cambium, for it is in this region that the separation occurs. Some cambial tissue adheres to the lifted rind, whilst some remains attached to the wood, and much of it is moist and slippery to the touch. This cambium, whether remaining on the wood or on the rind, is able to

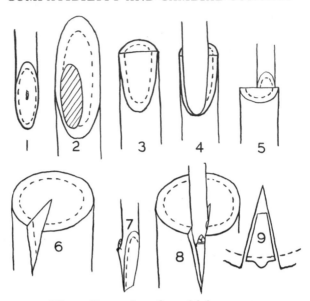

Fig. 4 Examples of cambial contact

1. Scion with thin rind. Cambium (dotted line) close to the outside of the rind. 2. Stock with thick rind. Shading indicates cut surface of scion and limits of cambial contact. 3. Stock prepared to achieve good apical and basal contact with scion cambium. 4. Scion applied to stock. Note good contact at base and matching of inner rind (cambium) rather than outer rind (bark). 5. Good cambial contact at top of stock. 6. Large stock with thick rind prepared for thin scion. 7. Scion with thin rind. 8. Stock and scion fitted. Note parts of stock rind outside the scion. 9. Cross-section. Note alignment of cambia and unmatched barks.

continue growth, laying down new rind on the outside and new wood on the inside, provided it is protected from the air and does not become infected with destructive organisms. It is of interest to note that rind lifted by winter frost separates beneath the cambium, so that growth occurs only from inside the lifted rind and not from the surface of the wood. Moreover, cambial tissue is less injured by frost than much other.

Wood surfaces exposed by frost-lifted rind may be healed by tissue development of the ends of the medullary rays, in the absence of a true cambium, such surfaces are prevented from drying as described under 'veneering with rind only' (page 171). Frost-lifted rind should be nailed or tied back into position and sealed with petroleum jelly, flexible wax or tape.

TISSUE CHANGES INDUCED BY GRAFTING

When stressing the two vital factors for success, compatibility and cambial contact—particularly the need to place cambiums in contact—the possible contribution made by transformation of adjacent living tissues should be recognized[94]. This contribution comes only from living cells, once again underlying the value of good craftsmanship and the maintenance of life in stock and scion. Even though cells have ceased to divide they may be reactivated to serve particular purposes. Under favourable conditions the severing of mature vascular tissues by incisions is followed by reconnection across regions where no procambium existed before wounding[127]. It appears that living plant cells can change from one form to another in response to surroundings or stimuli, hormonal or physical. Adventitious developments, as in the formation of sphaeroblasts and burr-knots, are common examples of such a change of function. Reversals of form and function, apparently not due to any recognized artificial cause, have been observed as, for example, when parts of white petals become green leaves or shoots become fruits and fruits become shoots (Plate 9).

PRESSURE WITHIN AND WITHOUT

Tissue character partly depends on physical pressure during development. The distinctly contrasting patterns of cellular differentiation obtained experimentally, with and without applied pressure, indicate the essential role played by physical confinement[11]. This is an important factor in our understanding of the healing of wounds and in grafting. An open wound permits unorganized cell proliferation to form a healing callus, but when this callus is confined and under pressure differentiation takes place in accordance with the various stresses governed by depth, orientation and proximity or congestion, emphasizing once again the need to provide close contact and firm anchorage in grafting operations: anchorage not only of tissue to tissue but also fixing in space relative to physical forces of gravity and wind, involving changes in pressure within tissues, followed by responsive structural development[98]. Whilst hereditary disposition

9. Change of form and function

A. From petal to green leaf in sweet cherry. B. Pear fruits into shoots and back again. C. Pear fruit with vegetative buds in spiral order. D. Two-year-old shoots of Prunus laurocerasus *producing thousands of roots following bacterial canker infection below.*

patterns growth, light, gravity, wind and other physical forces—soil anchorage and the mechanics of staking—determine development[45].

Tissue reactivation in response to environmental change—physical or chemical—entails consumption and a gradient or flow of materials towards such changing tissue[82]. The whole path of such movement is reactivated: the ridge which develops below a successful graft, or between a particular branch and root, exemplifies this behaviour.

CHAPTER III
Rootstocks and their Propagation

Most methods of grafting are designed for the propagation of selected varieties which are not so conveniently multipled in other ways. These varieties, used as scions, are generally grafted upon closely related plants which bear the roots and are known as rootstocks. This grafting does not directly increase the total number of plants but merely increases one variety at the expense of another.

Rootstocks have important effects upon the scions worked upon them and a great deal has been written concerning rootstock behaviour[59,60,61]. The attributes of the innumerable kinds of rootstocks cannot be evaluated here, but it is possible to give brief descriptions of the chief methods used in their propagation. These methods fall under two headings, namely propagation by seed (sexual) and by vegetative means (asexual).

ROOTSTOCKS FROM SEED

Rootstock propagation by means of seed is by far the most common method and has the great advantage over vegetative propagation that it entails rather less work. Seed is collected and sown, and, provided simple rules are observed, very little else is necessary to obtain more or less unlimited numbers of individuals which are usually healthy and vigorous—at least whilst they are young. These properties of seedlings are most valuable and should not be ignored when comparing seedlings with vegetatively-propagated material. The chief disadvantages of seedlings are their propensity for variation and the impossibility of predicting the limits of this variability.

The nature of the seed determines the technique employed,

but some general remarks which have wide application may be helpful to those who have not already learnt by experience.

SOURCES OF SEED

When a wide range of species is required, and the need varies from year to year, commercial seed houses generally prove the most convenient source. They handle a wide range of species from many locations at home and overseas. Costs vary widely and purchasers should obtain information concerning age, viability, germination, method of storage and guaranteed delivery date.

When regular supplies of native seed are required and such are available locally, it may prove less costly to collect direct from local parks, gardens or estates. Records should be kept of the location and behaviour of each batch of seed so that only the better sources are used in subsequent years. Seed should be taken only from healthy trees which are representative of the type required in relation to growth and form. Stunted trees tend to produce stunted progeny.

COLLECTING THE SEED

It is generally stated that seed should be well matured before it is harvested and this, no doubt, is the result of long experience, but still it sometimes happens that immature seed sown immediately after gathering will germinate more quickly than mature seed. Nevertheless, for practical purposes, maturity is advisable, especially where the seed is to be stored for some time before sowing.

The time of collection varies somewhat from year to year. Some seed, *Acer saccharinum* for example, must be collected in midsummer; cherry (*Prunus avium*) soon after; *Betula* spp. in late summer; *Crataegus*, *Fagus* and *Quercus* spp. in the autumn. Beech (*Fagus sylvatica*) requires careful examination to make sure that a good proportion is sound. In a good year ninety per cent may be viable but in a bad year most of the husks may be empty.

Some species—beech is an example—produce good seed crops only at intervals of several years and seed collected in the years of plenty must be stored for later sowings. The chief aim of storage is to maintain the best possible seed viability.

STORING

Seeds which are naturally dry when harvested, such as many of the conifers, are best kept so until sown. Cool, dry conditions are ideal but cones, or extracted seed, of many conifers retain satisfactory viability for some years in a dry, airy loft. When dry storage is not normally available the seeds are best kept in sealed containers. The majority of seeds, once they have been dried, will keep well in sealed containers at temperatures ranging from just above freezing to around 4°C (40°F). Seeds which are surrounded by succulent material, such as plums and cherries, should never be allowed to become thoroughly dry, but should be stored in moist conditions at comparatively low temperatures (*see* Stratification *below*). The quickest way to germinate apple seeds is to sow them immediately they are taken from the fruit, so that they have no time to dry. This is obviously not possible when the seed has to be transported long distances in quantity, and in commercial practice the seed is freed of fruit-pulp, partially dried, and mixed with an equal bulk of charcoal for dry storage.

In the tropics sour orange, mango, and avocado pear seeds are sown for rootstocks more or less immediately they have been freed from the pulp of the fruit.

Large, oily seeds such as those of *Aesculus, Castanea, Hevea, Juglans* and *Quercus* species are comparatively short-lived owing to drying and loss of food products under warm conditions. They should be sown soon after collection or stratified for six to twelve weeks in moist material before planting (*see* Stratification *below*). With nuts it is necessary to provide adequate protection against rodents.

Some seeds, such as those of the olive, are surrounded by oily pulp, and germination is delayed unless this is completely removed. Leaving the ripe fruit in heaps to rot, followed by washing, or, alternatively, feeding the fruit to animals, effectively disposes of the pulp. For quick germination clean olive seeds are cracked, or the pointed end is clipped off, prior to sowing.

On a small scale large, hard-shelled seeds can be individually cracked, either in a vice or with pincers, but care is necessary to avoid damage to the embryo. Other ways of increasing the permeability of the seed coat include rubbing between two sheets of sandpaper or tumbling with grit or gravel in a churn. The treatment must cease well before the seed is injured. Cracked or scarified seed, having lost some of its protective coat, keeps less well in store and is best sown soon after treatment.

Seed must be protected from mice and other animals, and where stored in large bulk it may be advisable to include some chemical to prevent deterioration. Naphthalene balls placed among stored seeds help to keep away weevils and other insects.

STRATIFICATION

Where seeds have to be kept moist between harvesting and sowing, or where dry seeds are given a pre-sowing treatment under cool conditions, they are placed in moist material, often in layers, and these treatments are known as methods of stratification. On a small scale, where special facilities are not available, the seed may be placed in plant pots between layers of moist sand or sand/peat mixtures (Fig. 5). These pots are usually buried completely in open ground. When it is necessary to separate a number of small lots of seed, each kind

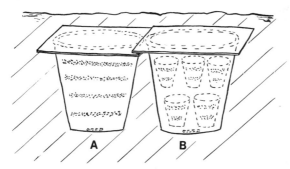

Fig. 5 Stratification of small lots of seed
A. Seed in alternate layers with moist sand. Pot and drainage hole covered with slate, or flat tile, and all buried about 150 mm (6 in.) underground in a moist situation. B. A few seeds of each of many kinds placed in small pots within a larger. Each small pot must be indelibly labelled.

may be placed in a small pot and a number of these within a large receptacle. The pots must be closely covered with tiles or slates to prevent the entry of mice. Small lots may also be placed in damp peat, or peat and sand, in small polyethylene bags in a home refrigerator.

Larger quantities of seed may be stratified in wooden boxes which are usually buried. The boxes should be completely covered with fine-mesh wire netting.

In nurseries where large quantities of seeds are given pre-sowing treatments, special methods and materials are used according to the nature of the seed and local conditions. A description of treatments advocated for seeds of apple, holly, plum, and rose (*Rosa canina*) will serve to indicate the conditions which lead to success with these different types.

Apple (Pyrus malus). The treatment begins some eight weeks before sowing. Air-dried seed is mixed with one-third its weight of charcoal and placed in small sacks, about 14 kg (30 lb.) in each so that the sacks are never more than half-filled. A large vessel of water is heated to 74°C (165°F). Each sack of seed is submerged and moved about in this heated water for ten seconds, then the sack of seed is removed and laid flat on slats to cool for thirty minutes, when the dipping process is repeated. On taking out of the heated water the second time the sacks are plunged, and moved about, in a tank of cold water and then put into a cold chamber maintained at about 2°C (36°F). The sacks of seed must remain wet, yet aerated. Laying out singly on wooden staging or shelves and frequent sprinkling of the walls and floor with water have the desired effect. The bags should be turned and shaken every day or so and watered thoroughly as required. After six or seven weeks the seeds begin to germinate and a close watch must then be kept on samples taken every three or four days and sown in warmth. The appearance of roots in these samples in four days indicates that the bulk of the seed is ready for sowing outdoors. At no time from the beginning of the treatment until they are sown must the seeds become dry.

An alternative treatment is to stratify the seed immediately after extraction, whilst still wet, and to sow it in the open in early spring. Still another method, suitable for small batches, is to extract the seed when the fruit is ripe in autumn, drop it into a vessel of water and swill occasionally over a period of two days, and then sow directly in pots or boxes, which are placed under glass. Given warmth, germination takes place in two or three weeks but otherwise is delayed till spring.

Holly (*Ilex aquifolia*). Holly is very slow to germinate and may require up to three years from sowing. The berries are collected in autumn or winter and either sown thickly in drills or stratified in moist material. If stratified, the seed is usually left so until the second spring, some fifteen months after collection, and then sown thickly in drills along with the stratifying medium in a shady moist situation. Germination occurs over a long period and the seed bed is left intact for two or even three years from sowing. When most of the seeds germinate only after two years' stratification it may prove more economical to discard the early crop to obtain a more even stand.

Plum (*Prunus* spp.). Stone fruits require about three months' pre-treatment and it is therefore necessary to start well in advance of sowing. The stones are distributed in rather more than their own volume of a mixture of one part sand to three parts pulverized peat. Small sacks are half-filled as described for apples, and given the same treatment except for the extra month or so in the cold chamber.

Rose (*Rosa canina*, *R. dumetorum*, and other spp.). Production of rose rootstocks from seed is a highly specialized business and is generally left to large-scale suppliers. Rose 'seed' (botanically an achene) is deeply dormant when harvested and requires a long period of chilling before it will germinate. In nature this takes at least two winters. Rootstock raisers have devised various ways of hastening germination, one more or less imitating nature, the other using sulphuric acid. In either case the ripe hips are hand-picked from established stock bushes when deep orange in colour, rather than when fully ripe and red. After crushing with rollers they are placed in tanks of water for two or three days, where they begin to ferment. After some vigorous stirring and repeated washing and draining to remove pulp and other debris, the seed is spread out to dry on old newspapers in a shed or greenhouse for two days. Each kilogramme contains about 66,000 seeds (30,000 per pound).

Normal eighteen-month stratification takes place in concrete bins or other receptacles erected in the open. These must be well drained and fitted with fine-mesh wire netting against birds and mice. The seed is mixed in moist sand or mixtures of moist sand, peat or vermiculite, and thereafter turned weekly for some eighteen months until germination begins. At the first signs of germination (chitting) the mixture is either immediately sown in wide drills in well-manured soil free of weeds or, alternatively, cold-stored until sown. The rate of sowing is important to the rootstock raiser. To achieve a good saleable grade of 5–8 mm (about $\frac{1}{4}$ in.), about 165 seeds should go to a metre (50 per foot), depending on seed quality and local conditions. The seed must not become dry at any time.

Germination has been improved by a warm period of moist stratification followed by a period of cold. Four weeks at 21°C (70°F) followed by three or four months at 5°C (40°F), prior to sowing, has given good results.

Sulphuric acid treatment[9] has greatly improved and hastened germination. A combination of acid treatment followed by the above-mentioned controlled temperature storage has given excellent results. To each kilogramme of dried seed placed in an acid-proof vessel add 625 ml of concentrated sulphuric acid ($\frac{1}{2}$ pint per pound), enough to wet the seed all over. Stir continuously with a glass rod for twenty to forty-five minutes, depending on the thickness of the pericarp (seed-coat). The period required is judged by the degree of blackening of the seed-coat, ascertained by cutting a sample seed and observing the progress of the blackening. To end the steeping period, dilute with a large volume of water; stir and change the water till all traces of acid are removed. Drain, dry and then rub off surface black by sieving, mix with equal volumes of moist sand and vermiculite, place in a nylon net bag and give controlled-temperature storage until time of sowing. There is considerable risk attending the use of the concentrated acid necessitating great care by the operator.

STRATIFICATION IN SITU

Large seeds such as those of peach and apricot are sometimes 'sown' close together in prepared beds which are kept moist and are protected from vermin. When germination occurs each seed is lifted and immediately planted in rows where the young plants are to be worked.

Another method is to drill the seeds in rows spaced for working and to thin them, leaving single plants 150 mm (6 in.) or more apart. In this way a large proportion of the seedlings is discarded, but those that remain grow rapidly, unchecked by transplanting.

SOWING

Land intended for raising rootstocks from seed should be clear of perennial weeds and fallowed at least one year before sowing, during which it is thoroughly cleaned. Deep, rich, friable, and well-drained soils are best, as these encourage good rapid growth of the young rootstocks.

Where the seed ground is horse- or tractor-cultivated the seed is sown in single rows spaced for this purpose. Large seeds such as walnut and peach are usually set individually about 150 mm (6 in.) apart in the rows. Smaller seeds, such as apple and pear, are sown much thicker in the rows or broadcast in prepared seed beds. These seed beds are often about 1·2 m (4 ft.) wide with 0·6 m (2 ft.) paths between. When broadcast in beds the seed must be covered with sand or other non-capping media. When sown in rows the seed should be covered 25–50 mm (1–2 in.) on either side. Rates of sowing must be governed by the nature and germination propensity. Seedling development largely depends on sowing rate.

With walnuts and many other seeds, it is essential to protect against birds. On a small scale this is conveniently done by placing netting over the beds, but on a really large scale bird-scarers must be in use from dawn to dusk.

TRANSPLANTING

There is a wide variation in the rate of growth of various seedlings. In general only the most rapid growth will enable transplanting to take place after one season, and two years in the seed bed is more usual. Seedlings left for a second year are commonly undercut after one year to encourage fibrous root development. The beds may be cleared at one time, or only the best rootstocks withdrawn and the rest left to gain size the second year. First-size seedling rootstocks may be planted, or potted, for grafting. Smaller sizes are bedded a short distance apart in nursery rows for one or two seasons before transferring to the place of grafting.

At transplanting time the roots are trimmed with a knife to facilitate planting. The prepared rootstocks may be dipped or fumigated at this time according to their particular needs.

VEGETATIVE ROOTSTOCK PROPAGATION

The chief advantage of the vegetatively propagated rootstock over the seedling is uniformity. The efficiency of plants propagated vegetatively is rarely questioned and the advantages of the practice have long been recognized. The unpredictable variability of seedling rootstocks will probably not much longer be tolerated. Determination to persist in vegetative propagation has brought into use very many different methods too numerous to describe individually here. They can be conveniently considered under two headings, rootstocks from cuttings and rootstocks from layers[40].

Rootstocks from Cuttings

Cuttings are parts of plants which are separated from the parent and treated in various ways to encourage the production of a complete plant. The cutting may be a piece of root, stem, leaf or merely a single bud or eye. Under specially-controlled conditions exploiting laboratory techniques of tissue culture described as *in vitro* or micropropagation, using a tiny section of apical meristem, it has been possible to obtain very high multiplication, even up to a millionfold per year. Work at East Malling, England[73] has made possible the production of many thousands of apple, plum and cherry rootstock plants from a single shoot-tip within one year.

If the cutting is a piece of root, then this must grow a shoot from an adventitious bud and continue to extend in order to maintain a balance between shoot and root. Similarly a shoot cutting must produce adventitious roots.

Cuttings must be placed in conditions where they are prevented from drying yet have ready access to air. Thus a good nursery soil is open in texture yet moist and easily worked. Artificial media made to many recipes are used throughout the world according to local conditions and the nature of the cuttings.

The source of the cutting is of considerable importance[40], for cuttings of one variety will regenerate in differing degree according to the condition of the parent plant and the part used as a cutting. The behaviour of particular varieties must be ascertained by experiment.

THE USE OF HORMONE SUBSTANCES

The cuttings, once taken, can be treated in various ways to hasten regeneration. Hormone-containing substances, often termed growth substances, have been used with considerable, but varying, success. These substances are sold under various proprietary names and should be used as directed by the vendors. The pure substances may be obtained from chemical firms. Two of the most widely used for promoting root growth on stems are β-indolyl-butyric acid, often known as indole butyric acid, and α-naphthalene-acetic acid. Of the various ways in which growth substances are applied the most practicable are the concentrated dip method and the dust method.

The concentrated dip method. The bases of stem cuttings are momentarily dipped, a handful at a time, 5 mm (just under $\frac{1}{4}$ in.) in an alcoholic solution of the hormone and planted when dry. To prepare, dissolve 5 g of the substance in 500 ml ($\frac{1}{4}$ oz. per pint) of ethyl alcohol, methylated spirit or acetone, and dilute with an equal volume of water to give a solution of 5,000 parts per million. The pure substances can be obtained in convenient 1 g tubes. Weaker 'dips' are obtained by diluting with more alcohol and water; stronger dips may prove injurious. Stored in well-stoppered opaque bottles, in a cool place, the stock solution retains its activity for about a year.

77

The dust method. The bases of the cuttings, a number at a time, are wetted by dipping into 50 per cent alcohol or 50 per cent acetone and, after shaking off excess liquid, into dust so that the lower 5 mm (just under ¼ in.) is well covered with dust. The cutting is immediately planted in a trench or by means of a dibber, though no deeper than usual. Pharmaceutical talc is the usual carrier, but charcoal and other materials have been used with success. Dusts, ready for use, may be purchased or can be made as follows. The growth substance is dissolved in sufficient alcohol or acetone to make a pasty mixture when stirred into the talc. This mixture is occasionally stirred whilst drying in a fan circulation of unheated air in a darkened room. To prepare 100 g (4 oz.) of dust containing about 5,000 p.p.m., dissolve 0·5 g (1/50 oz.) of the growth substance in 40 ml (1½ fl. oz.) of alcohol, stir into 100 g (4 oz.) of talc and dry as described. This dust keeps for months if stored in a closed container in a cool, dark place.

SHOOT CUTTINGS

Each of several factors which influence the regeneration of plants from cuttings has at some time been regarded as the most vital. Some workers have considered the nature of the cutting itself as supremely important; others the soil or rooting medium; or skilled preparation and nursing; or the use of growth substance stimulants; or the provision of artificial environments. All of these, as well as others not mentioned, or even realized, can be important; but the individual factor operates only in relation to the remainder and all must be considered together. For cutting behaviour is determined not only by variety and source, by preparation and hormonal treatment, by adjustment of environment including medium, humidity, light, and temperature, but is profoundly affected by their interactions.

Shoot cuttings may be taken at different stages: soft (herbaceous), firm (semi-lignified), or hard (woody). Cuttings from young shoots in rapid growth root quickly provided they are carefully nursed, for being highly sensitive and quick to root they are likewise quick to die. When the shoot is fully grown it is relatively tough and insensitive and at this hardwood stage the range of plants which root is more limited than at the softwood stage. The semi-lignified cutting comes between the soft- and the hard-wood cutting; too sensitive for planting in the open, it requires limited protection for some time.

Hardwood cuttings of some species, notably of currant, poplar, privet, prunus, quince and willow, are expected to take care of themselves and to survive whatever the elements provide; and being independent of expensive glasshouses, or other protection, may be of any desired initial size. Nevertheless, even ready rooters will do better with some judicious culture. Usually one-year-old shoots are cut into pieces 150–450 mm (6–18 in.) long and these are firmly planted at least half their length underground. Larger cuttings make larger plants, provided they root readily.

With ready-rooting subjects almost any part of the shoot may be used with success, but often better results are obtained by careful selection of cuttings. Basal parts of yearling shoots are usually best, and the presence of a piece of two-year wood is helpful with some subjects. Thus cuttings may be straight or heeled. Cuttings with large heels or 'mallet' bases are sometimes used, but are not beneficial with all species. Various kinds of base are depicted in Fig. 6.

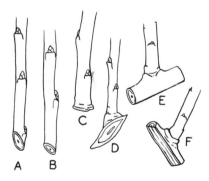

Fig. 6 Straight and heeled cuttings
A. Straight cutting with bud close to its base. B. Straight cutting with base between buds. An 'internodal' cutting. C. Straight basal cutting. Cut at the junction with older shoot. D. Cutting with heel of older wood. E. Mallet cutting. F. Cleft-mallet cutting.

79

In temperate climates hardwood cuttings of some species are best taken and planted in the autumn whilst bearing at least a few leaves[40]. Under these conditions the technique for handling 'straight' cuttings is as follows. Ripe shoots, still bearing leaves, are cut into 300 mm (12 in.) lengths, the soft upper part of the shoot being discarded. The cuts are made regardless of buds and it is usual to cut handfuls of shoots in one movement of knife or secateurs. The cuttings are never allowed to wilt and are planted the same day with any adhering leaves. Rootstock cuttings are often planted about two inches apart in rows spaced to permit hand or machine cultivation, though an efficient herbicide treatment at planting is better than mechanical cultivation. It permits less precise alignment, obviates hoe damage and prevents weed competition. The soil should be moist before planting begins; otherwise the bed should be well irrigated.

Cuttings are planted vertically, or nearly so, with some three-quarters of their length, or 175 mm (7 in.), underground. A spade is thrust into the soil to make a 'grip', or cleft, into which the cuttings are inserted. The cuttings are made as firm as possible at their base by treading. The planting is completed by levelling off the soil. If winter frosts lift the cuttings they should be pushed down and well firmed by treading.

Recent research, mainly with fruit tree rootstocks, has revealed that established hedges provide better shoot cuttings than nursery plants. Nursery plants are not only comparatively poor sources of cuttings but there is grave risk of virus contamination if they have been grafted with unhealthy scion varieties. Hedges, once established true to name, reduce the risk of admixtures, a common fault with cuttings collected from a changing nursery population.

Hardwood cuttings should be taken in the autumn, given an appropriate growth substance treatment and then planted directly into the soil if the subject is easy to root, or leafs-out early in spring.

With subjects that are somewhat slower to root from autumn cuttings it is worthwhile taking them as the leaf-fall is finishing, giving them a growth substance, bedding them in a well-aerated medium and providing protection against cold

10. Winter storage of hardwood cuttings

Straw bales and thatched hurdles conserve warmth necessary for the initiation of roots prior to planting.

for two or three months prior to planting (Plate 10). If the temperature at the base of the cutting is prevented from falling much below 7°C (45°F) considerable root initiation and, in warm winters, root development may occur. With shy rooters, as are most apple rootstocks and many forest and ornamental trees, additional warmth is essential to ensure root initiation in the dormant season.

Warm storage treatment. In order that roots develop quickly and planting take place before shoot growth begins, warmth is applied to the base of the cutting. This is particularly important when cuttings are collected in late winter towards the end of their dormancy. The rate of rooting increases with rising temperature to an optimum at 21°C (70°F). Cuttings collected in late winter (February) require only two or three weeks' bottom heat treatment; longer periods will exhaust carbohydrates and curtail establishment. Heat is applied to

81

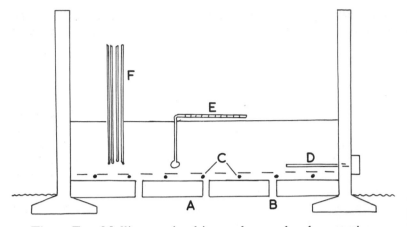

Fig. 7 East Malling cutting bins under overhead protection
Vertical section (diagrammatic) after Misc. Pub. 85, East Malling Research Station. A. Coarse drainage. B. Drainage between floor blocks. C. Warming wires attached to underside of 'Weldmesh' screen. D. Rod thermostat. E. Earth thermometer. F. Cuttings.

the base of the cuttings only, by electrical soil-warming wires with a loading of up to 160 watts per square metre (15 watts per square foot), controlled by a thermostat. This equipment is placed in an insulated bin (Fig. 7), preferably sited in a cool, insulated building such as a north-facing open-sided stone barn or penthouse. The cuttings are not bundled in the stores but are set thickly in rows alternating with about 25 mm (1 in.) of the storage medium of equal parts of coarse sand and granulated peat. In this way one thousand hardwood cuttings can be accommodated in about 0·3 m² (a little less than 4 sq. ft.). Detailed instructions and further up-to-date information resulting from continuing 'warm storage' investigations are obtainable from East Malling Research Station, Maidstone, Kent, England.

Propagation by hardwood cuttings is simple and economical, but not all subjects respond. Pear (*Pyrus communis*) rootstocks in general root but little and cherry (*Prunus avium*) fails completely. Rose rootstocks of a number of species respond very well; to reduce suckering propensities rose cuttings are 'eyed' (disbudded), leaving only the upper two or three buds.

Practically all one-year, rooted, hardwood cuttings are fit to line-out for working the following season. (*See also* page 222.)

SOFTWOOD CUTTINGS

Most rootstocks can be propagated by means of herbaceous or softwood cuttings struck under controlled conditions in propagation frames. The source of the cutting is important and cuttings taken early in the growing season are superior to late cuttings for the practical reason that the rooted cutting has time to become thoroughly established before the winter. Lateral shoots, not actively elongating, are preferred. The ideal material, being early and yet firm, is produced on bushy stock plants established some years in pots and brought into early growth in a warm house.

Terminal growths some 75 mm (3 in.) in length are collected and placed in baskets lined with moist material. In the propagation house the bases are cut internodally. Careful cutting when collecting makes this second cut unnecessary. No leaves are removed. Various hormone treatments, as previously described, may be used to hasten rooting.

The cuttings are planted in clean quartz sand or, better still, in a mixture of one part sand to two or three parts moist, granulated peat. A bottom heat of 21–24°C (70–75°F) encourages rooting.

The sand or other medium is firmed, levelled off, and watered. All water is obtained from a tank within the house, so that it is warmed before use and does not seriously reduce the temperature of the medium, as it would if used direct from the main supply. The cuttings are set upright, just deep enough to prevent them falling over when watered. No dibbers are used, except when dust treatment has been given, and each row is watered in from the side across the surface of the medium so that this is firmed about the base of the cuttings. The distance between the cuttings is governed by the size of the foliage: 50 mm (2 in.) between the cuttings and 75 mm (3 in.) between the rows is usually adequate spacing.

Watering and light-shading is given to prevent wilting, but the maximum amount of light, other than direct sunlight, should be admitted to keep the leaves active. The lights are raised each morning and wiped free of condensed moisture, and any rotting leaves are removed at this time.

Rooting often begins some ten days after planting and the bulk are fit to pot into 50 mm (2 in.) pots in twenty days from setting. The potting mixture should be open and contain no fresh dung which might cause loss by rotting. Equal parts of fibrous loam and sharp sand suit most woody subjects. A small quantity of old rotted dung may be placed in the bottom of the pot for drainage and to provide a source of nutrient later. Pots and compost should be placed in the house overnight to warm up. The rooted cutting is placed at the side of the pot and the compost gently firmed against it. The potted cuttings are replaced in the case, or in a case at a similar temperature. In a few days air is admitted and the potted cuttings pass to the open bench, then to unheated frames outside the house and, finally, into nursery rows where they are planted 100–150 mm (4–6 in.) apart. The next season they grow large enough to line out for budding, two years after setting in the cutting bed.

Mist propagation. The constant skilled attention required for softwood cuttings in closed cases under glass has somewhat limited the use of the softwood cutting method, particularly where there is a shortage of skilled operators, or where labour costs are comparatively high. Early attempts to replace the traditional closed-case technique were largely frustrated by the lack of reliable automatic devices. Excellent humidifiers, sprays and mists are now procurable and there is widespread interest in their use.

Mist propagation as we know it today is mechanization of the frequent hand syringing long practiced by gardeners[4]. The system has been well described[114] as mechanical and controlled syringing to maintain a film of water on the leaves at all times. This results in a reduction of leaf temperature leading to reduced transpiration and respiration, enabling propagation under higher light intensity with accompanying high rates of photosynthesis. The combination of these factors makes possible the use of soft leafy cuttings of the highest regenerative capacity, which would otherwise fail to survive.

Mist is a superlative nurse and therein lies a danger. In this environment the cutting becomes acclimatized and then, when mist is no longer required, the cutting needs to be gradually hardened to a normal environment. This is the age-old horticultural problem; the more deeply we nurse the material the greater the hardening problem. So we should avoid superfluous mist application and reduce it at the earliest possible moment that the newly-rooted cutting can stand alone.

There are many systems: one of the most successful provides a mist spray of a few seconds at intervals governed by a hydrostat, commonly known as an 'electronic leaf'. A reliable supply of clean water at a pressure above $1 \cdot 75$ kg/cm^2 (25 lb./sq. in.) is needed and fine nozzles of the deflection or baffle type designed to minimize clogging. The interval between sprays, and their duration, is controlled by the position of the hydrostat in relation to the nozzles and also by adjusting the electrical resistance in the relay mechanism.

It is essential that the rooting medium be very well drained or waterlogging may occur. Many mediums have been tried; fine gravel or coarse sand, mixtures of sand and peat, vermiculite and, in the tropics, composted sawdust over coarse drainage[31].

Mist systems may be used in glasshouses, plastic tents, or in the open. It is necessary to screen the area from air currents which may blow the mist away from the cuttings. This is not usually necessary under glass but is essential in open situations.

The speed and degree of rooting of leafy cuttings under mist are governed by the nature of the material and the temperature of the medium. In unheated systems in the open in England many woody subjects require some eight weeks for rooting, but this period has been shortened by the use of bottom heat. In warmer climates, unheated, open-topped bins, with or without moderate shade, have given excellent results.

Material rooted under mist must be hardened to withstand drier environments. This is done by a gradual reduction of the mist *in situ* or by transference to glasshouse or other protection. Cuttings leaving the mist in autumn do best if overwintered under glass or plastic tunnels. Ready-rooting herbaceous cuttings have been greatly helped by mist environments supplementing the normal horticultural method involving protection and bottom heat. Cuttings may be taken at a softer, more sensitive stage of growth, more light can be admitted to the cuttings, and fungus diseases and insect pests are greatly reduced.

Cuttings in cold frames. Many subjects, including evergreens, hardy and half-hardy shrubs, and trees, are readily propagated by means of half-ripened cuttings placed in a sandy medium in unheated frames. Cloches, hand-lights or plastic tunnels may take the place of frames. The season for this work extends from the end of July until the following spring. In general, lateral shoots make the best cuttings and these may vary in length from 50–200 mm (2–8 in.) according to the nature of the subject. Some of the finer-leaved conifers and heaths may be no more than 25 mm (1 in.) long.

Firm planting in a sand and peat medium in non-fluctuating conditions of temperature and moisture are the keynotes to success. Protection from excessive light is best given by slatted blinds or screens, or by open-meshed cloth such as coarse hessian, preferably supported on a light frame at least a few inches above the frames. In cold weather the frames must be closely covered with mats, especially at night. Home-made straw or rush mats, made by simple weaving, are ideal for this purpose. In very cold weather, straw packed around and between ranks of frames will assist the maintenance of an equable temperature.

Cuttings taken in late summer will often begin to root before winter; those taken in the autumn may callus during the early part of the winter and begin to root in the spring. When the cuttings are rooted, air may be gradually admitted more freely until the lights are removed altogether. Some light-shading, preferably by lath screens, may be necessary through the summer. The established cuttings are fit to line-out in the open ground the following autumn at distances suited to their growth.

Cuttings under double protection. The Dutch 'double-glass' method provides the essential environment with very little attention throughout the rooting period. An unheated frame on the ground, with sun-facing slope from 450–300 mm (18–12 in.), has an internal horizontal frame of 450 × 25 mm (18 × 1 in.) boards to take a Dutch light, which must be absolutely flat; otherwise inside condensation will run off, leaving the glass clear. The cuttings are heavily watered and immediately enclosed under the close-fitting Dutch light. The sloping light above must be shaded in all but dull weather. Complete misting of the inner light indicates that watering can be postponed. When the cuttings are well rooted the inner light is lifted by stages and watering begins. As the cuttings grow the light is removed; the sloping light is left as a protection against spring frosts. This system is particularly good for evergreens.

One of the simplest practical ways to propagate somewhat ready-rooting cuttings, particularly conifers, hollies and the like, is as follows. On a well-drained site of weed-free soil spread a 75 mm (3 in.) layer of rooting medium within sleeper or other heavy boards spaced to support glass lights at a slight slope to the sun. Set the cuttings, water heavily and lay on them a sheet of very thin polyethylene. If the edge of this is fixed to a batten it can be adjusted and prevented from wind damage. The sloping glass light should be about 150 mm (6 in.) above the plastic cover. Watering is only necessary at long intervals. When rooting is established the plastic is removed gradually, and when weather permits, the glass light also. Slat shade is sometimes used to prolong the protection. The plants having rooted into the soil below may be left for a season before transference.

LEAF-BUD CUTTINGS

A limited number of shrubs and trees are fairly readily propagated by leaf-bud cuttings and, where successful, this proves a very rapid means of multiplying valuable specimens. Leaf-bud propagation of numbers of the *Rubus* family has been used on a commercial scale. Success in these cases has been correlated with the capacity of the species to root by tip layering.

Apples have responded in varying degree, the best results

coming by the use of material in the juvenile phase. Shoots arising from root pieces—adventitiously—have rooted well, but those from stems, in the mature phase, hardly at all[116].

Only the current season's vegetative shoots provide suitable material. The shoots are collected when the majority of the leaves are fully expanded, that is at the height of the growing season, and are kept moist until the cuttings are made. All the leaves are used except the small unexpanded ones at the tip. The shoot is held by the lower end and a shallow cut (A) is made 12 mm ($\frac{1}{2}$ in.) below the leaf, as in Fig. 8. The knife is removed, and a second cut (B), beginning 12 mm ($\frac{1}{2}$ in.) above the leaf, passes under the bud (C) to meet the first cut (A). This second cut should not penetrate deeply but rather as described later for shield budding. The cuttings are allowed to fall into a suitably placed receptacle, containing damp moss or cloth, to prevent them drying.

The cuttings are planted in a mixture of sand and peat, or a similar medium. The use of bottom heat in closed cases increases the speed of rooting. The cuttings should be planted

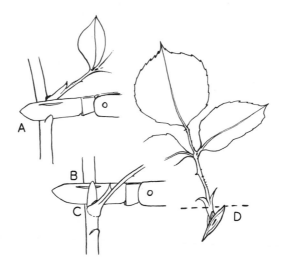

Fig. 8 Leaf-bud cuttings

First a shallow cut (A), joined by a larger cut (B), which removes a thin slice of wood and rind with the bud (C) and the leaf. The cutting is planted no deeper than indicated by the dotted line (D).

as quickly as possible after preparation. The base of the cutting is set so that the bud and small portion of stem attached to it is just below the surface of the medium and the whole of the upper surface of the leaf is exposed to the light. Wilting is prevented by attention to shading and watering.

When rooted, the cuttings are boxed or potted and gradually hardened off as described for softwood cuttings.

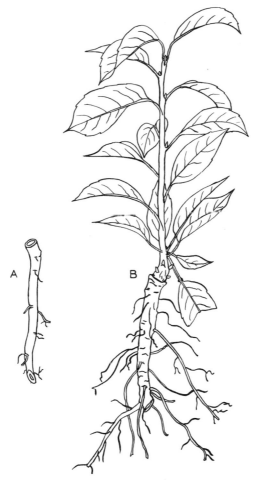

Fig. 9 Propagation by root cuttings

A. Cutting of sweet cherry (P. avium) *as planted in early spring. B. The same the following autumn.*

ROOT CUTTINGS

Many rootstocks, shrubs, and trees are fairly readily propagated by root cuttings, provided young roots are used. Two-year roots regenerate much more readily than those four or five years old. Apple, pear, plum, and cherry rootstocks are often raised by root cuttings.

When trees are lifted from the nursery the roots are trimmed and these trimmings, 6–20 mm ($\frac{1}{4}$–$\frac{3}{4}$ in.) in diameter, are cut into lengths of 100–150 mm (4–6 in.). To make sure of planting the cuttings the right way up, the end nearest the crown of the plant is cut square across and the other slantwise. (Fig. 9.)

It is usual to tie the cuttings in bundles of twenty-five and to cover them with moist soil or sand until planted. In late winter or early spring the cuttings are planted vertically in well-dug ground, 50–75 mm (2–3 in.) apart, in rows spaced for hand cultivation. The upper end of the cuttings is just covered with soil. If subsequent rain washes away the soil so that the upper 3 mm ($\frac{1}{8}$ in.) of the cutting is exposed, so much the better. Should late frosts lift the cuttings, they must be pressed down again and well firmed. A number of shoots grow from each cutting, and when 25–50 mm (1–2 in.) high, they are carefully reduced to one. A piece of cleft bamboo of pencil length may be used to facilitate the separation and thinning of the shoots.

Rootstocks from Layers

The term layering is used to cover all processes in which a part of the plant, usually the stem, is induced to grow roots and/or shoots before separation from the parent. There are various practical methods, the chief of which are stooling, layering, and all types of marcottage (air layering).

STOOLING

This is the simplest layering method. The parent plant, having first become established, is cut down to the ground and resulting growths from the stub (stool) are earthed up. (Fig. 10.)

The parent rootstocks are planted upright, 0·3 m (1 ft.) apart, in rows 1·1 m (3 ft. 6 in.) apart. Beyond the usual clean

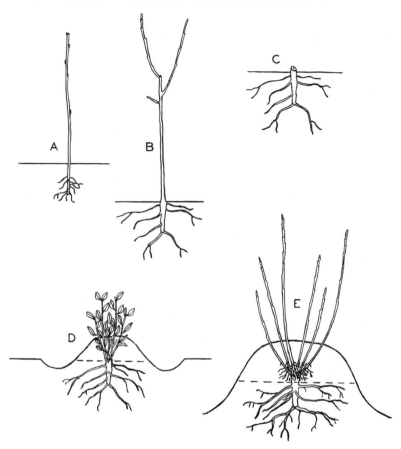

Fig. 10 The stooling method

A. Newly-planted parent. B. Established parent. C. Cut down one year after planting. D. Earthed a first time in early summer. E. Rooted shoots on the stool in winter.

cultivation no special treatment is given during the first season.

The established rootstocks are cut down close to the ground during the winter following that in which they were planted. This hard cutting results in the production of a number of shoots, and these are earthed to half their height when they are 125–150 mm (5–6 in.) high. Earthing is repeated after further growth, but the shoots are at no time buried more than half their total height. At each earthing the soil is worked amongst the shoots so that all light is excluded from the bases of the shoots. The final earthing brings the soil 150–200 mm (6–8 in.) above the base of the shoots.

At the end of the growing season, when the plants are dormant, the soil is removed and all the shoots, whether rooted or not, are cut or broken away from the stool. Established stools must be completely unearthed; otherwise it is not possible to remove the shoots close to the stool and, in a few years, they become so raised by the addition of shoot bases that it is impossible to earth them. After harvesting, the stools are left exposed until a further crop of shoots arise, when these are earthed as in the previous year. Given good treatment, stools remain productive for twenty years or more. Well-rooted shoots are fit to line-out for working and the sparsely rooted should be planted thickly in nursery rows for one year, rather like hardwood cuttings, only not so deeply.

LAYERING—ETIOLATION METHOD

Layering, or pegging down of shoots, is not so simple as stooling and is therefore reserved for the comparatively shy-rooters. Of the various techniques employed, the etiolation method is the most reliable and practical. A very wide range of plants will root by this method. Apples, pears, quinces, plums, cherries, peaches, walnuts, mulberries, and various economic tropical plants, as well as many ornamental trees and shrubs, have responded satisfactorily.

The technique is designed to bring about the etiolation of the base of the young shoots. The parent rootstocks are planted when dormant at an angle of 45° along the row, some 0·6 m (2 ft.) apart. The rows should be widely spaced to facilitate earthing, 1·2–1·5 m (4–5 ft.) usually being sufficient.

The soil is kept well cultivated and the parent layers are pegged down in the following winter. Late winter is preferred to autumn because it decreases the risk of frost lifting the pegs and undoing the work. Any weak laterals are cut back hard, and strong laterals are lightly tipped. Along each row a trench is made about 50 mm (2 in.) deep and wide enough to receive the layer without cramping. The parent layer is pegged down

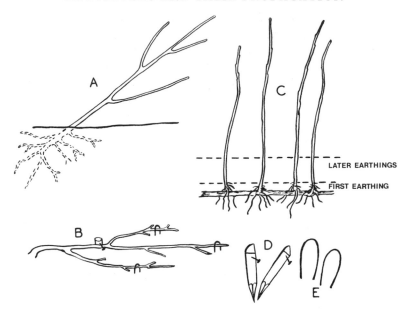

Fig. 11 The etiolation method of layering

A. Newly planted parent. B. Established and pegged down one year after planting. C. Rooted shoots on the parent layer one year later. D. Home-made wooden pegs with 50 mm (2 in.) wire nail near upper end. E. Wire (10-gauge) doubled for pegging down slender shoots.

flat in this shallow trench. A little soil is removed near the base to facilitate this operation. The layer is held down by pegs made from sharpened pieces of wood, 250–300 mm (10–12 in.) long, having a 50 mm (2 in.) nail at right angles about 25 mm (1 in.) from the upper end. Small shoots may be held by wire pegs made by doubling 380 mm (15 in.) lengths of 10-gauge wire in the form of a letter U (Fig. 11). Any shoots not held down flat must be cut off close to the ground.

The entire layer is covered with about 25 mm (1 in.) of friable soil, or peat-sand mixture, before the buds swell, so that all new growths must push through this layer and are completely etiolated. Before these young shoots have un-curled and expanded their leaves, a further 25 mm (1 in.) of soil is placed upon them. No soil is added to areas where shoots have not appeared. This is repeated two or three times during the first few weeks of the growing season. It has the effect of ensuring the etiolation of the first 25 mm (1 in.) or so of the young shoot and it is from this area that roots will eventually appear. Without this etiolation there is little or no rooting.

After the initial earthings, and when the shoots have grown 75–100 mm (3–4 in.) above the soil, earth is again added to half the height of the shoot. Further additions are made at intervals of a few weeks until the bases of the shoots are buried 150–200 mm (6–8 in.).

The second winter following planting the soil is forked away and the one-year-old rooted shoots are removed at their junction with the parent layer. Unrooted shoots are left to be pegged down as at first. If all the shoots are rooted, some of the most vigorous must be left for pegging down; one strong shoot per 300 mm (1 ft.) of row is sufficient. At each layering the laid shoots must be kept as flat as possible.

Once harvested, the crop is handled as described under stooling.

MARCOTTING

This method is variously known as air layering, Chinese layering, circumposition, or, in India and elsewhere, as gootee. Essentially the treatment is to injure or girdle a shoot and surround the wound with moist material until roots are put forth, when the rooted shoot is severed from the parent and established as a new individual.

There are various ways of making the wound, but the most efficacious is to remove a complete ring of rind, up to 25 mm (1 in.) wide, from a shoot which is carrying some well-developed leaves. In the northern temperate zone the season is from early April to the end of August. There is no need to place the girdle in any particular relation to buds. The rind should be removed completely from the ring so that there is no connection outside the wood of the shoot. In this operation the wood must not be cut, for this would weaken the shoot and cause the marcot to break. The girdled part is then packed with a ball of compost and an outer wrapping of sphagnum moss or other moisture-holding material which must be kept moist. This is conveniently done by a wrapping of polyethylene film sold under various proprietary names such as 'Alathon', 'Alkathene', 'Visqueen', etc. These materials are available in sheets, or as lay-flat tubing of varying width and thickness. A suitable thickness for marcots is 75- or 100-gauge (0·003 in. or 0·004 in.). The shoot or young branch having been girdled is painted with an alcoholic solution of growth substance as used for 'quick-dip' treatment of hardwood cuttings (*see* page 77). A handful of moistened moss or compost, squeezed free of water, is placed around and a little above the ring and is then wrapped in film (Fig. 12). The film

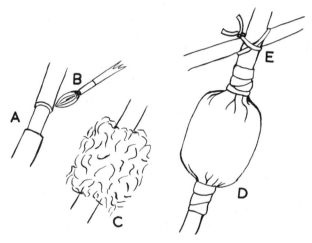

Fig. 12 Marcotting

A. Shoot ringed. B. Painted for 25 mm (1 in.) above the ring with growth substance. C. Moss. D. Film enclosure. E. Supported by a neighbouring branch.

is tightly folded and secured to the branch with a number of turns of adhesive transparent cellulose tape such as 'Sellotape' or 'Speedfix'. A piece of film 150 × 200 mm (6 × 8 in.) is suitable for a marcot of average size. Care is taken to interfold the edges to prevent the entry of rain. Lay-flat tubing, 100 mm (4 in.) wide, has proved more convenient, provided it can be passed over the upper part of the shoot. The treated branch is supported by tying its upper part to a neighbouring branch.

Rooting is normally well developed by the end of one growing season, and the marcots are then separated from the parent plant. The film and surplus root-free moss are removed, the new plant is potted in an open medium, and is then put in a closed case or cold frame till well established. Unlike stooling and the etiolation method, in marcotting a separate operation is required for each individual propagated and therefore the method does not lend itself to unlimited extension; it is, however, a very sure method of propagation and has rarely been known to fail.

OTHER METHODS OF LAYERING

Tip layering, where young branches are bent down almost vertically and buried shallowly in the soil, is a useful method for some subjects, notably certain *Rubus* and *Ribes* species. Serpentine layering, or the intermittent burying of long shoots, usually accompanied by notching the parts buried, and so-called simple layering, where the shoot is bent to rest on the soil and is there earthed, leaving the tip of the shoot exposed, are other methods frequently used. However, none of these has such wide application as either the etiolation or the stooling method of layering.

Grading Rootstocks

The amount of roots on a rootstock cannot alone determine the grade, as roots cannot be considered apart from size. A medium-sized rootstock well rooted is superior to a large specimen that is poorly rooted. The quality of subsequent growth is important and a rapidly-growing rootstock is better for working than one which has been slow to get away.

Length of Time between Planting and Working

Where rootstocks are grafted before planting, as in bench grafting, the question of time for the rootstock to become established does not arise, but it is an important factor where rootstocks are to be worked in the nursery row. A rootstock planted in the winter is usually budded in the following summer, when in rapid growth. The bud normally remains dormant until the following spring when, the part of the rootstock above the bud having been removed, the dormant bud grows out to form the maiden tree. If the rootstock is grafted with dormant scions the cutting-down of the rootstock is done in the second spring after planting. It is most inadvisable to graft young rootstocks the first spring, within a month or two of planting. The resulting trees are usually very weak and fail to keep pace with trees worked the second spring.

A newly-planted rootstock has far more stem than root, and new root growth is very rapid to restore the balance. When such rootstocks are cut hard in the spring following budding, or at grafting, the roots then exceed the stem and the new growth from the scion is vigorous and results in a quickly-grown maiden tree.

On the other hand, if a newly-planted rootstock is cut hard and grafted, the effect of surplus stem is lost and root growth is comparatively meagre. Furthermore, there is no stage at which the hard cutting of an established rootstock can occur and consequently at no time is rapid, straight growth encouraged.

CHAPTER IV

The Collection and Treatment
of Scion-wood

Trueness to Name

Seeing that the scion maintains its characteristics after grafting, it is important to procure scions from plants which have been correctly identified. Thus it is worthwhile having all parent plants carefully labelled or otherwise marked. When more than one variety of scion is collected, each separate bundle or container should be carefully labelled at that time.

With some subjects, notably those in which mutations (bud sports) have occurred, it is essential to select scion material only from those parts of the plant which exhibit the characters desired. This calls for marking in the flowering or fruiting stages or at such other times as the desired characters are clearly revealed.

Numerous examples of mutation or 'sporting' are found in citrus, and the importance of recognizing this phenomenon so that scions are taken only from the more desirable strains has been stressed[108]. Careful observations revealed that many orchards, presumed of one citrus variety, contained a number of trees differing from the normal accepted type. Where these alien characteristics persisted in the progeny raised by budding, the trees were designated as different 'strains'; the word 'strain'[108] being used 'to designate a group of individuals of a horticultural variety differing from all other individuals of the variety in one or more constant and recognizable characteristics and capable of perpetuation through vegetative propagation'. It is obvious that bud mutation must sometimes show itself in a superior instead of an inferior characteristic, resulting in improved forms.

Some subjects vary comparatively little in vegetative characters, yet the fruit may be different. An example of this is seen in the cherries, both sweet and acid, and here it is highly important to mark the parent trees whilst fruiting. This precaution obviates the continual propagation of wrongly-named varieties, as happens all too frequently when scions are saved year after year from successive 'generations' of young, non-bearing trees. The selection and marking of individual scion-source trees must be followed, at least every few years, by a review of progeny behaviour, to avoid perpetuating an undesirable mutation or systemic disease only visible in the flower, fruit or mature wood.

Virus Diseases
Careful choice of plants from which scions are collected is vital if the risk of spreading virus diseases is to be reduced.

E. Yoxall Jones   Copyright East Malling Res. Stat.

11. Graft failure due to virus-infected scions

Ten rootstocks (left) were worked with symptomless Early Rivers cherry, later shown to be virus infected, and only four grew. Ten (right) worked exactly similarly, but with healthy scions, all grew into good trees.

The parent plants should be observed closely during growth and those selected for propagation should be given an identification number, which is repeated when labelling the daughter scions. In this way the behaviour of the new tree will provide further information on the suitability of the parent as a source of scions. Where several parent trees are used the propagator can considerably reduce virus diseases by discarding parent trees which produce abnormal offspring; the time involved in recording is a very small addition to the normal nursery procedure. The simplest method[103] is to have a number for each parent plant, adding this to the variety name on the scion labels. So, instead of a mixed batch of scions there is, for example, 'Early Red' 1, 2, and 3, etc., and the progeny performance can then guide the choice of scions in the future resulting in improvement in quality and, probably, in grafting successes, as shown in Plate 11. The improvement of scion varieties in relation to virus disease involves the use of virus-free stocks; the same careful selection for health must, naturally, apply to each component.

Labelling

When scion-wood is to be stored for any considerable time the labels and markings should be durable. Storage under moist conditions calls for rot-proof labels and attachments. Non-corrosive metal has proved excellent for most purposes, provided the wire or other connection is equally lasting. Certain proprietary plastic materials are also satisfactory, provided the connections are rot-proof. Painted wooden labels, affixed with pliable wire, are quite satisfactory for most purposes.

Labels should be attached so that they cannot slip along and off the end of the scion material. Where bundles are tied with two 'strings' the label should be fixed between these, or, where the material is branched, between one side branch and another. With some stout, straight scion material, not bundled, the label may be threaded through a hole in one of the pieces or a slice may be cut from the side and the exposed wood marked with waterproof pencil or with paint. Colour coding provides a means of easy recognition, achieved by dipping the shoot tips in paint, or wrapping individual shoots with a single turn of adhesive tape.

Even the best methods of labelling should be supplemented by notes in the garden- or nursery-book.

Herbaceous and Leafy Scions

Leafy scion material loses water rapidly unless kept in a humid atmosphere, and for this reason it is impracticable to store such material successfully. It therefore becomes necessary to graft herbaceous scions as soon as possible after collection; meanwhile they may be kept in a fresh condition in containers lined with moist material such as sphagnum moss, or sealed in polyethylene bags and kept in a cool, shady place.

The leaf area may be reduced by cutting away some of the leaves, or, better still, by reducing the size of each leaf, immediately the shoots have been collected.

Scions for Bud-grafting (Budding)

For summer budding, shoots are collected, at the time suitable for the operation, from healthy parent plants of the chosen variety. It is usual to remove the leaves immediately to reduce water loss. This is done by severing the petiole about 12 mm ($\frac{1}{2}$ in.) from its base. The soft tip of the shoot and the stipules may be removed at this time. These defoliated shoots, or bud sticks as they are called, may be kept in good condition for some days by standing them in a pail containing about 25 mm (1 in.) of clean water. Both pail and bud sticks should be covered with a damp cloth and kept in a cool place. They also keep in perfect condition for many days if closely wrapped or sealed in polyethylene film, provided not more than 1 kg (2·2 lb.) of scions is placed in each parcel and it is kept in a cool, dark place. The petiole bases absciss after a few days, but this is more a sign of viability than detriment, save that the budder may prefer to handle petioled material.

For transport over really long distances full use should be made of air transport. The weight of the consignment may be kept down by reducing the length of the bud sticks to contain only the best buds. These are first wrapped closely or heat sealed in thin 50-gauge (0·002 in.) polyethylene, neither wet material nor internal packing of any kind being used. The wrapped scions are then covered with an outer wrapping for protection against bruising and to receive the address label and any necessary permits.

UNPETIOLED BUD-WOOD

The petiole base adhering to the bud stick serves as a convenient 'handle' in the budding process, but it appears to serve no other useful purpose. In subjects where the base of the petiole is large it may prove a disadvantage and should be removed. If the petioles are severed midway some ten to fourteen days before collecting the bud sticks, the petiole bases will absciss, leaving a healed scar which cannot serve as a point of entry for disease. Close cutting hastens abscission. With some subjects there is no difference in the 'take' (success) of buds whether the petioles are removed at the time of collection or in advance[88].

PRE-GIRDLED SCIONS

Some subjects, notably in the tropics, may be so active in growth that they soon exhaust their food reserves when separated from the parent plant. Others may not normally have sufficient reserves in their shoots at the grafting season. In these cases grafting success has been considerably improved by girdling the future scion at its base while the leaves are fully active so that the necessary materials accumulate in the future scion. A knife-edge ring one month before collecting the scion may serve, but the removal of a complete ring of rind, without injuring the wood, is more definitely effective.

In Malaya buds of Durian (*Durio zibethinus*) taken from branches previously ringbarked at their base some ten days before, took and grew better than those from non-ringed branches.

DORMANT BUD-WOOD

For spring budding, before the current season's shoots have developed buds in the leaf-axils, use is made of the previous year's production. The bud sticks are collected whilst dormant and are stored in a cool, moist situation. Storage in moist material in a cool chamber at 0–2°C (32–34°F) is ideal, but the sticks keep almost as well packed in moist sphagnum moss in boxes buried underground. Many subjects remain in good condition, if bundled and heeled-in on the shady side of a building. Alternatively they may be enclosed in polyethylene sheet and stood in a really cool cellar. If the spring budders wish to remove the wood from the bud shield of autumn-collected sticks, this must be collected before the leaves fall, defoliated, sealed and stored at 0°C (32°F).

Scions for Dormant Grafting

For grafting in the late winter and early spring, scion-wood is collected whilst completely dormant, tied tightly in small bundles with shoot bases level. The bundles should be no larger than a man's handful; larger bundles dry out in their middles. In severe climates it is necessary to place the bundles in storehouses, or cellars, or to bury them completely underground. Normally, however, the bundled scions keep excellently if placed upright with their bases some 150 mm (6 in.) in soil or sand on the shady side of a building or thick hedge. The site should be protected from drying winds, but the heeled-in bundles should be open to the sky. With open-stored scion-wood it is essential to keep the bundles strictly upright, a little spaced and open to the sky, to reduce the risk of rodent damage.

Disinfection of Scion-wood

It is not always possible to guarantee that only disease-free material is used for scions, and infected material may, in certain circumstances, be disinfected and temporarily protected against reinfection by treatment with appropriate chemicals. Insects, such as many aphides, may be killed by immersing the scion-wood in nicotine and soap solution at concentrations recommended for spraying intact plants in a similar stage of growth: 15 g ($\frac{1}{2}$ oz.) of 98 per cent nicotine in 40–50 litres (8–10 gallons) of water is usually effective and safe.

There is considerable evidence of the transfer of certain diseases from the scion-wood trees to the nursery in the form of latent spores, invisible to the naked eye until they cause lesions in the nursery material. Treatment with fungicide has given a high degree of control of disease transfer[79,7], commercially available fungicides based on captan (Orthocide ®)

didecyldimethyl ammonium bromide, DDAB (Deciquam ®) dodine (Cyprex ®, Melprex ®) and ferbam (Fermate ®) having been used at conventional concentrations. Both bud-wood and scion-wood should be treated, preferably by dipping in an aqueous dispersion of the fungicide and allowed to become completely dry before grafting operations begin. Working with rubber (*Hevea brasiliensis*) it was found to be advantageous to treat also the area on the rootstock destined to receive the bud patch. This was conveniently done by wiping with a cloth dipped in the fungicide dispersion. Whilst the above treatments provide a high degree of protection, it is only by completely freeing the scion source plants from infection that the highest possible standard of control is attained.

Potassium permanganate solution 'of a dark red colour when viewed in a glass tumbler' has been used as a dip for scion-wood of pawpaw (*Carica papaya*) to prevent rotting and also as a sterilizing wash applied to the surface of the stock before grafting[57]. The actual cut surfaces placed in contact were not treated in the experiment cited.

Storage of Prepared Scions

In some forms of grafting, notably those involving skilful or complicated preparation of the scion, some nurserymen set one or more skilled men to cut the scions, which are later grafted to the stocks. Provided pre-cut scions are prevented from drying, and are kept cool, they will remain viable for some weeks, during which they may be transported over long distances. Additional protection against drying is obtained by momentarily immersing the graft-wood, when cut into scion length, in a quick-drying anti-desiccant (*see* page 217) such as a diluted vinyl polymer emulsion, or a dissolved natural resin which once set does not smear. The coated scion piece may then be prepared, leaving a minimal evaporating surface.

CHAPTER V

Tools and Accessories

With a sharp knife, a light billhook, an ordinary saw, some string, and non-poisonous sealing material, an enthusiast may successfully accomplish almost all types of grafting, but by the use of tools which have been well proved, the work is made easier and quicker. Of the tools employed knives come first in importance. Notes are given below of various edge tools, tying and sealing materials, and other accessories.

Knives

There are six kinds to be considered, first the general grafting knife; then the budding knife; the two-edged (Mexican) knife; the surgical knife; and 'double-bladed' knife; and finally the pruning or trimming knife. Knives with folding blades are more convenient to carry and the cutting edges are protected against damage when not in use, but fixed blades are stronger and are particularly valuable for heavy cutting and trimming and for preparing large hard-wooded stocks for grafting. Small and delicate knives such as scalpels or lancets, and razors, are commonly made with fixed handles and are best kept in their special cases between operations. The joints of all folding knives should be oiled to lessen wear, which is otherwise the main cause of their deterioration.

THE GENERAL GRAFTING KNIFE

This should be of good quality and have a straight-edged blade strongly set in a handle which is large enough to afford a comfortable grip. Cheap knives quickly become loose and useless. A good knife for general work is shown (Fig. 13, F and G). The blade of this knife is about 75 mm (3 in.) long, set in a handle of 100–125 mm (4–5 in.). An inexpensive knife for general grafting is made in France; it has a folding blade which is locked when in use.

THE BUDDING KNIFE

Specially made for normal shield budding, it is also used for many kinds of grafting where the subjects are delicately constructed or are somewhat herbaceous. The chief patterns are illustrated (Fig. 13). The first, knife A, is a light one commonly used in Britain. The total blade length does not exceed 50 mm (2 in.), whilst the handle may be as long as 100 mm (6 in.). A cutting edge which curves away to the tip is usually preferred because this makes it easier to 'flick up' the stock rind at the finish of the upward cut in T shield budding. Sharply-pointed knives tend to enter the wood of the stock. At the opposite end of the knife, the bone handle is extended and thinned down to form a spatula which is used for lifting the rind.

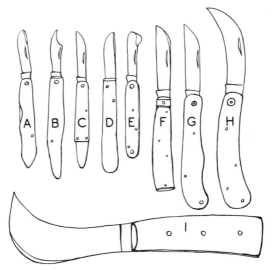

Fig. 13 Knives

A. A lightweight budding knife popular in England. B. and C. Budding knives common in continental Europe. D. Inexpensive budding knife with fixed handle. E. Budding knife with protuberance on the blade for lifting rind. F. English grafting knife, useful for general work. G. Grafting knife much used on the Continent. H. Trimming knife with folding handle. I. Heavy trimming knife with fixed handle.

Knives B and C are in common use in continental Europe. The handle of C has a thin spatula, of bone or metal, fitted within a thicker handle. Knife D is a useful and inexpensive budding knife with a fixed handle, common in North America, whilst knife E has a substantial handle; a protuberance on the back of the blades serves to lift the rind.

Whichever type of knife is selected it should be light in weight and well-balanced, engendering daintiness in the operator.

THE TWO-EDGED (MEXICAN) KNIFE
This is readily made from a table knife by grinding. There are two sharpened edges, along one side and across the chisel-like end (Fig. 14, A and B). The side (A) is used normally but the chisel end (B) is pressed vertically through the rind without any sliding. It is specially useful when budding with an inclined incision (page 147). An adept with this knife, because of the reduced number of hand movements compared with conventional budding, can obtain very high speeds, but speed is secondary to good workmanship (page 230).

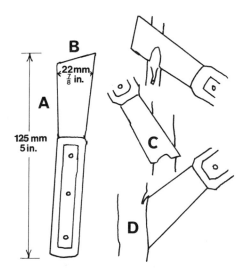

Fig. 14 Two-edged (Mexican) knife
Side A used normally. Side B pressed vertically through the rind at C and D without sliding.

107

THE SURGICAL KNIFE

For the grafting of soft herbaceous plants the blade of the knife must be thin, as well as sharp, to avoid all risk of crushing the tissue. 'Cut-throat'-type razors are usually much too thick for good work. Safety razor blades are useful but should be fitted with a handle. The ideal implement for really delicate tissues is the modern surgeon's (Bard Parker) knife with detachable blade. These handles and blades are of various shapes and sizes and it is possible to select a suitable knife for the object in view.

THE DOUBLE-BLADED KNIFE

For patch budding the double-bladed knife is almost essential. Some operators make use of two of these implements, one to remove the rind from the stock and another, slightly larger, to cut the scion. This ensures a tight fit at the edges of the graft in spite of the shrinkage of the scion and the contraction of the stock rind.

Well-made implements may be bought. Some have folding blades and some are fixed. Home-made tools may be constructed by fitting two knife blades or scalpels to a small block of wood. Safety-razor blades make excellent implements if well set into wooden or metal handles. The construction

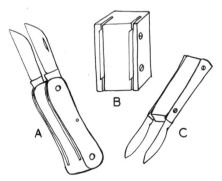

Fig. 15 Double-bladed knives for patch budding

A. A useful folding knife. B. Two safety-razor blades placed in narrow saw-cuts in a block of wood. Adjustable screws pass through the blades and clamp the wood. C. Two all-metal dissecting scalpels bored and screwed to a piece of wood make a useful implement.

should permit easy exchange of the blades when worn or broken. A useful light handle has been devised to make use of razor blades in this way. Some typical implements are illustrated (Fig. 15).

THE PRUNING OR TRIMMING KNIFE
The grafter should preserve his grafting knives for the actual operation and have additional knives of heavier construction for preparing the materials, especially those which are hard-wooded. These knives will also serve for cutting the binding materials and for all those operations which tend to injure the cutting edge. A generally useful knife is the ordinary curved-bladed pruning knife; samples of both folding and fixed types are depicted in Fig. 13, H and I.

SHARPENING THE KNIFE
A grafting knife should be supplied with a smooth, sharp edge strong enough to withstand the strain imposed by twisting or curving cuts. The usual straight knife recommended for general grafting work can be perfectly sharpened on a flat stone. A moderately coarse-surfaced stone may be used on a new knife, but thereafter only a finely-textured hone should be used. The stone, whether 'oil' or 'water', should remain fixed throughout the sharpening operation and *not* be held in the hand. The whole width of the stone should be used so that its surface remains flat. The knife is held to the stone at an angle of between 20° and 25° and moved forward along the stone against the cutting edge. The edge is thus 'pushed' on to the knife and not drawn away by the stone. When beginning to sharpen, the direction of the movements may be changed every half-dozen strokes but the operation should be finished off by single strokes in alternate directions. A correctly sharpened knife will retain a good edge for many days' work and during it, only requires stropping on a leather, or boot, to retain a perfect edge.

The above remarks apply to knives with blades sharpened equally on both sides. Knives sharpened by bevelling on one side will be retained so by keeping one side flat on the hone, to suit either left- or right-handed workers. Such 'one-sided' knives tend to 'bite' one way and 'skid' the other, but they are excellent for making perfectly flat cuts, as required for the veneer side graft.

The whole blade should be kept clear of dirt and congealed plant juices by scraping and polishing, as necessary.

Cleaving Tools and Mallets

Small stocks are conveniently cleft, or split, by the knife, but for large stocks it is customary to use special tools. For cleaving along the diameter of large stocks any stout blade will serve. An ordinary billhook may be modified by a blacksmith to form a useful and everlasting cleaving tool. In some quarters a blade with a curved edge is advocated. The object of the curved blade is to cut, rather than tear, the rind (Fig. 16).

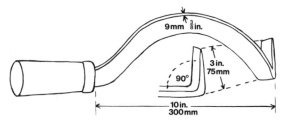

Fig. 16 A curved cleaving tool

Straight cleaving tools tend to split and tear the stock rind. A curved blade cuts the rind ahead of splitting the wood.

For oblique cleft grafting a smaller implement is necessary. Often these tools are too large and the blade too thick for good work, and—for this reason—a scale drawing is provided (Fig. 17). The use of the tool is described on page 111.

Where cleaving tools are not provided with wedging devices, for holding open the cleft, metal or wooden wedges must be used.

The cleaving tools are usually driven by mallets of hardwood. Oak, or one of the very hard woods produced in the tropics, is suitable. A club or truncheon shape is ideal. Experienced cleft-grafters usually thread a 0·6 m (2 ft.) thong through the handle of the mallet and attach it to the cleaving tool, so that the mallet dangles from the inserted tool, and the hands are left free to cut and insert the scions.

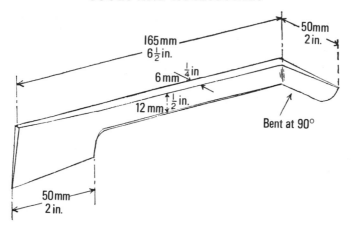

Fig. 17 Tool for oblique cleft grafting

Drawn to scale to guide the blacksmith or toolmaker. The sizes given should not be exceeded.

The Combined Grafter

Ancient French books[3] have described a tool used for inlay grafting. One of these tools is illustrated (Fig. 18), but it seems doubtful if it merits a place in the grafting tool-box, largely because inlay methods are comparatively rarely used in modern times.

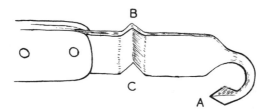

Fig. 18 Combined grafter for inlaying

A. A gouge for forming an angular channel in the stock. B. A V-shaped blade to form the scion to fit into the channel in the stock. The tool shown may be sharpened at B and C to suit both right- and left-handed operators.

The Inlaying Chisel[26]

This is used when inlaying large woody stocks (*see* 'kerf grafting' described on page 166) and will also be found generally useful for trimming any large clefts.

The chisel (Fig. 19) may be specially made or adapted from a heavy quality domestic mincing knife. Set at an angle in the handle, for ease of manipulation, the blade should be kept 'knife-sharp' and used for preparing the scion as well as the stock. For shaping the scion, the hand grasping the tool is held still by steadying it against the knee and the scion is drawn against the motionless tool. Tools of an over-all length of 200–225 mm (8–9 in.) are a convenient size.

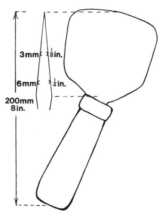

Fig. 19 Inlaying chisel

This may be adapted from a domestic mincing knife of heavy pattern. The blade should be well sharpened and used for preparing the scion as well as the stock.

Grafting Machines

When stock and scion are of equal shape and size at the point of grafting, as is the yearling wood of vines, various cutting guides, slicers, and planes may be adapted to accelerate the output of unskilled workers. One of the simplest 'machines' is merely a short steel tube with one end bevelled to correspond with the slanting cut required on the stock and scion. The parts to be grafted are pushed through the tube in turn and the bevelled end serves to guide the knife as it makes the slanting cut. More complicated machines are those in which the shoots are held by a pincer, and a hinged blade moves across at a set angle.

These contrivances, with many modifications, are in use for the bench grafting of vines in Italy, Spain, and elsewhere.

Electrically driven circular saws have been used to mortice and tenon the ends of stock and scion, so that they hold together without tying. Then there are double-acting foot-lever machines which cut scion and stock in one movement; one of the best is made by Ulysse Fabre, Vaison-Vaucluse, France. A hand-operated horizontal guillotine is made by A. Lozevis, Place Pelletan, Agen, Lot-et-Garonne, France. This slices and tongues in one movement; the blades are conveniently removed for sharpening and can be adjusted as they wear.

Grafting machines are more commonly used for vines than for other subjects. Vine buds are widely spaced and in between the shoots are very regular, clean and hard, rather like dowel rods and quite unlike the majority of woody plants. Success with vines, therefore, does not necessarily indicate that other plants can be handled equally well. All these machines except for certain portable kinds working on the secateur principle, are mainly of use in bench grafting.

Secateurs

An infinite variety of patterns has been offered by manufacturers, but none has proved equal to a well-sharpened knife for making clean cuts. Nevertheless, a pair of secateurs is almost indispensable for the preparation of rootstocks and trees for grafting, particularly for frameworking mature trees. Some pruning shears, with two cutting edges, though excellent for pruning, will be found unsuitable for close cutting of side shoots in congested branch systems.

Well-made shears are now available which closely simulate a knife cut. They have one curved blade which passes an anvil, scissor fashion, and are thus able to cut laterals cleanly from a main stem. With this type it is necessary to invert the shears when changing to the other side of the stem. The blunt-nosed type, in which the blade closes on the middle of an anvil, has proved generally useful in spite of a tendency to crush the rind on one side of the shoot.

Saws

The blade should be narrow so that it can pass between close-spaced branches, for the complete removal of limbs. The blade being narrow, it must obtain its strength by thickness of metal. The saw may be straight or curved and set for drawing instead of thrusting. These modifications are a matter of individual preference.

The rough or ridged surface left by the saw is usually shaved off with a knife. There is little evidence that this shaving treatment is an advantage, but where ends of limbs are left unsealed the rough surface certainly holds more moisture and may assist the establishment of certain diseases which enter through wounds.

When removing lateral branches the saw cut should be made flush with the surface of the supporting branch or trunk; similarly, when lowering the main stem the cut should be made close to a lateral branch and no 'snag' or stub left which, otherwise, invariably dies back, preventing rapid healing. Small limbs should be supported, as the cut nears its end, to prevent tearing the rind below the cut. Larger limbs should be removed by first undercutting them some distance from their base. The stub is then removed by a single cut close to the supporting limb.

Disinfection of Tools

The disinfection of grafting tools is sometimes essential to prevent the spread of disease from one plant to another. The choice of chemicals is strictly limited by the susceptibility of cut surfaces of plants to damage. Domestic sterilants based on sodium hypochlorite are safe and satisfactory. Under laboratory conditions the number of tools, such as knives and razors, is doubled and whilst one of a pair is in use the other is immersed in the solution. Fresh cotton wool is used each time to wipe the instruments dry before use.

Graft Seals and Wound Dressings

Cut surfaces exposed at grafting time are covered to prevent the entry of water and to limit the passage of air. They should adhere well and remain stable, at least until the union is complete, while at the same time allowing for changes due to growth. They must not injure the stock or scion either directly by their nature, as some spiritous solvents or metallic salts may, or indirectly by their colour, affecting the temperature of the plant tissues. Black or very dark seals may become too hot in sunny climates. Good sealing may help to prevent the entry of disease organisms and damage by predators, particularly when the material contains a disinfectant or insecticide. Very many materials meet these requirements and it is not surprising that a host of substances has been used. When seeking new ones, additional features should receive attention; ready availability, cost, ease of application and mode of setting. This latter point is important. For example, hot waxes set from within as they reach the cool surface of the plant, thus they do not creep. Cold materials, which set by evaporation (drying) such as dissolved materials or emulsions, set from without and may creep between the graft surfaces, preventing union. From early times various clay pugs have been used with excellent results, but in modern times, when speed in operation is regarded as most important, the so-called grafting-waxes and bitumen seals have become popular. A few examples from each class will indicate their special attributes and serve as a guide to the student.

GRAFTING-CLAY OR PUG

These bulky materials are only suitable for large grafts as used when topworking established trees or inarching or approach grafting stout stems able to support the weight of the pug. A suitable grafting-pug may be made by kneading two parts clay with one of fresh cow-dung. The incorporation of some cow-hair or other fine fibre prevents the material from falling away when dried. Alternatively the pugged graft may be wrapped with cloth or plant fibres of various kinds to hold it in place. When the graft has taken, the growth itself may slough off the dried pug, but obstinate pieces may be dispersed by striking them simultaneously on either side with wooden billets.

Plain clay, or pug, is useful for smearing into large cracks formed when cleft grafting large stocks. This prevents the entry of soft sealing materials which otherwise might prevent the joining of the plant tissues. Plasticine, as used for modelling, serves the same purpose as clay but is relatively more expensive.

HOT GRAFTING-WAXES

These 'waxes', melted and applied by brush, are probably the most popular graft seals throughout the world. They are reasonably cheap to make, easily applied, and fairly stable even in hot weather. Each formula may be varied to suit particular conditions. In the warmer climates the admixture of inert solid should be increased to prevent running at high temperatures, though for really hot climates the bitumen emulsions will prove more suitable.

A hot wax, well tested in Britain, is made from the following ingredients in the proportions given, by weight:

Resin	10 parts
	(say 2 kg or 5 lb.)
Burgundy pitch	$3\frac{1}{4}$ parts
Tallow	$2\frac{1}{2}$ parts
Paraffin wax	$2\frac{1}{2}$ parts
Venetian red	3 parts

First melt the resin and Burgundy pitch, then add the paraffin wax and four-fifths of the tallow. Melt all together, stirring all the time. When the whole is melted and just boiling *take the vessel off the fire* and stir in the Venetian red. If the Venetian red is well warmed before stirring into the melted wax, the air content of the powder will be much reduced and there will be less risk of the material effervescing and overflowing. Have a vessel of cold water at hand. *Well grease the hands with the spare tallow*; then pour about half a litre or a pint of the mixture into the vessel of cold water and pull it about as though making toffee. The hot wax quickly cools on the outside of the mass, but inside the wax remains molten for some time and caution is necessary to avoid burning the hands. Deposit each portion on a sheet of paper well smeared with tallow. When set hard place the cakes in a cool situation until required.

Dark-coloured seals become very hot in strong sunlight and are not so suitable as those of lighter colour[20]. Some bright red waxes have been more attacked by birds than waxes of nondescript colour. Black pitch waxes, which melt only at a high temperature and quickly become unworkable as they cool, must be continually reheated and are therefore more wasteful of time than those workable at a more moderate temperature which cool relatively slowly. Moreover, in cold weather a pitch wax threads badly between brush and graft.

A useful light-coloured wax based on resin, and workable at a moderate temperature, is made as follows:

Resin	10 parts
Beeswax	2 parts
Siliceous earth (fuller's earth)	1 part

Melt together the resin and beeswax and stir in the siliceous earth. To reduce any tendency to flow in hot weather the amount of inert solid may be increased.

A less expensive seal, based on paraffin wax, is made as follows:

Paraffin wax	5 parts
Siliceous earth	3 parts
Zinc oxide	1 part

Melt the wax and stir in the siliceous earth and zinc oxide. Stir in also a little turpentine to soften the mixture, as may be found necessary for local conditions.

These and similar waxes are applied, after reheating, by means of a brush, but they can also be applied by means of the finger. For finger application the wax must not be too hot and the finger should first be dipped in a mixture of soft soap and water carried in a pannikin attached to the side of the portable heating apparatus. Finger application, though uncommon, is rapid and results in a very neat job.

COLD BRUSH WAXES

These usually depend upon the evaporation of a volatile ingredient. They must remain in sealed containers until used and only limited quantities must be exposed at one time, during application. Here is a typical formula:

Resin	8 parts
Beeswax	4 parts
Talc	1 part
Methylated spirit	2 parts

Melt the resin, then add the beeswax. When melted and stirred remove from the heat and add talc, previously warmed, stirring all the time. When thoroughly mixed, remove the vessel to the open air away from any fire or naked light, and whilst warm, but not hot, add the methylated spirit gradually, stirring all the time. Store the mixture in air-tight containers in a cool place. The wax is applied with a small, fairly stiff brush.

RUBBER LATEX

Vulcanized rubber suspended in a colloidal state, a white liquid rather like thin cream, has proved harmless to plant tissues and makes an excellent and non-restricting graft seal. Various modified emulsions, normally used for mending leaks in synthetic roofing material, have also been used with success. These proprietary latex emulsions have proved excellent for coating large superficial wounds, as commonly inflicted on street trees, and for 'double-treating' the surfaces exposed when really large limbs are removed from trees. The cambial region and sap-wood are first sealed with the emulsion and then the middle of the wound, or dead heart-wood, is treated with a wood preservative.

HAND MASTICS

Hand mastics and waxes are quite satisfactory for use on a small scale but are generally found to be too time-consuming for large scale use. They must be readily pliable yet not so soft as to flow away from the graft under the influence of the noonday sun. The best hand waxes contain a fair proportion of beeswax, which is expensive. A common practice is to melt varying amounts of resin, beeswax and tallow together, stir and then pour this mixture into a bucket of cold water to permit thorough kneading and pulling. The hands should be well greased with some spare tallow to prevent sticking. Recipes vary in accordance with local weather conditions, the proportions ranging from 10 to 12 per cent tallow, 30 to 70 per cent resin, and 20 to 60 per cent beeswax. A small increase in the amount of tallow will soften the mastic; increasing the resin will harden it, whilst increasing the beeswax will improve malleable quality whilst maintaining stability. Mastics for hand application should remain firm when not in use and be softened for application by frequent kneading. Where the grafter works alone there is some danger of his hands carrying the seal on to the cut surfaces. With two workers this may be avoided.

BITUMEN EMULSIONS

These are now available specially prepared for dressing tree wounds and for sealing grafts. They are particularly useful for dressing wounds, as they exhibit excellent covering power and are quite harmless to trees. Where the method of grafting entails leaving gaps and crevices, as in some cleft methods, these emulsions are not entirely successful in covering the gaps and it is quite often necessary to coat over these gaps more than once before the seal is complete. Bitumen emulsions set in the presence of air and it is necessary to avoid exposing the liquid or paste in bulk. Before setting, the emulsion may be thinned with water, as desired, and care must be taken to wash out brushes at intervals and when ceasing work. Unless definitely known to be harmless to trees, bitumastic paints and waterproofing materials should not be used indiscriminately, as many of these contain materials which quickly kill plant tissues.

Bitumen emulsions require no heating and do not run in hot weather. Moreover, they are brushed on and therefore convenient to apply. Amongst the proprietary materials of this class are 'Arbrex', and 'Flintkote' or 'Tree Heal'. Being water emulsions, they may be thinned for ease of application but this should be minimal as dilution increases their tendency to creep (*see* page 115).

PARAFFIN WAX

A melted candle may serve to seal a few grafts in the home garden, but the wax tends to flake off unless applied really hot. When hot enough for brush application it is thin and 'watery', tending to run into cavities rather than seal them, and is therefore unsuitable for general use unless first ameliorated with other ingredients.

Plain paraffin wax is excellent for sealing well-fitted bench grafts (page 208). The wax should be kept liquid by melting in a water bath, as in a glue pot. The completed graft is given an instantaneous dip, head first, into the molten wax up to the base of the scion and then into cold water.

PETROLEUM JELLY

A heavy, crude form of petroleum jelly has been used for sealing grafts. This jelly is cheap and easily applied by the hands. It remains fairly stable and it is not seriously affected by normal climatic changes. In extremely hot weather there is some danger of the jelly melting and running down the stock. There are several grades of petroleum jelly and it is essential that only the heavy, crude forms are used, as lighter forms have been known to melt and cause losses amongst grafts.

Petroleum jelly remains in a more or less tacky state for several years and has proved troublesome in sticking to the clothes of workers.

Soft materials, such as petroleum jelly, have not proved so satisfactory for sealing untied cleft grafts, such as the stub and the side graft, as hard-setting materials. These last cement the scion in its position and help to prevent dislodgement.

Petroleum jelly has been used to prevent damage by red bud borer (*Thomasiniana oculiperda Rubs.*) to newly inserted apple and rose buds. The jelly is smeared with the finger over the tied bud so as to cover completely the bud and all cuts in the bark of the stock; 450 g (1 lb.) of jelly is usually sufficient for over a thousand buddings.

There has been some evidence[48] that treatment of newly-inserted apple buds, even in the absence of insect damage, has improved the stand in seasons of poor 'takes'.

WAXED CLOTH AND ADHESIVE TAPE

Prepared cloth or tape serves both for tying and sealing and is particularly useful in patch budding. Unless it is to be removed and used again it is advisable to use weak cloth which will burst as the stock swells. For the same reason the tape should not be bound layer upon layer as this will cause constriction. On the other hand, it is usually necessary to encircle the stock so that the tape laps upon itself and firmly adheres. Old cotton sheeting or calico dipped in grafting-wax is admirable and may be torn to any size required. In

Trinidad cotton tape is dipped into melted beeswax. Before use it is drawn through the fingers, which makes it more adhesive. It can be used more than once.

Cloth may be impregnated in various ways. The cloth may be rolled on to a stick before dipping in molten wax or it may be unrolled from one spool on to another, via the molten wax and a scraper to remove surplus wax. Where a large number of graftings require the same length of tape, the tape may be cut into lengths, rolled on small spools of a hundred or so lengths, and then dipped in very hot melted wax. For ease of unrolling the strips from the spool, the end of one strip may be lapped one inch over the beginning of the next, so that removing a strip raises the end of another.

Waxed cloth patches are sometimes used to seal bud grafts. These can be made at home by folding cloth strip or wide tape to and fro to form squares. A large nail or skewer is now driven through the middle of the squares and the folding cloth is trussed with string. The two folded edges are cut off with a sharp knife to free the squares, and whilst trussed, the whole is dipped in hot wax until well saturated. The prepared squares are lifted from the skewer as required and placed over the inserted buds. When the squares are too small to overlap round the stock, they must be tied down with raffia or twine.

Many hot waxes are suitable. Beeswax is one of the best. Another is made with 4 parts beeswax (by weight), 2 parts resin, 1 part tallow. Yet another and cheaper recipe is 4 parts resin, 1 part tallow, 1 part beeswax, and 1 part raw linseed oil. The *'boiled'* or treated linseed oil used for paints contains poisonous substances and should never be used in any graft seal.

Sticky plastic tapes are used with success by some nurserymen in lieu of both tying and sealing materials. A suitable dispenser is needed and the tapes should not be wider than 20 mm ($\frac{3}{4}$ in.); otherwise they cannot conveniently be wrapped spirally round the stock. These watertight bandages appear to encourage a good development of callus.

Wax Heaters

These are almost as varied as the materials they are designed to heat. For large-scale work in the open, two main types are suitable. The first is simple but practical. A container such as a 5-litre (1-gallon) paint tin holds the wax in sufficient bulk if half-filled. The handle should be refixed through the sides of the vessel immediately beneath the rim. Handles are usually soldered on and this soon melts when the pot is heated. Old paint vessels may be conveniently cleaned by burning them out upside-down over a wood fire, provided the vessel is removed immediately the paint has come away and before the solder melts. A hot fire is required for quickly reheating the wax, and this may be obtained by burning coke or charcoal in a pail, or small oil drum, in the sides of which numerous holes have been made for ventilation.

The other system is to have light, portable heaters combined with the wax pot. These may be purchased or made at home of metal or wood. The heat is usually obtained by burning alcohol or oil in a small lamp in the lower part of the

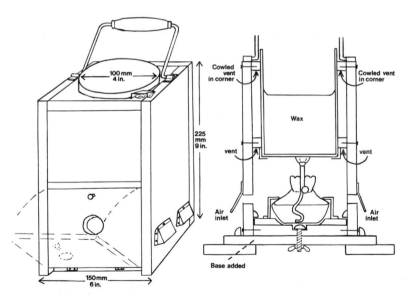

Fig. 20 Portable wax heater

An aluminium litre or quart measure and an inexpensive 'night-light' (to burn methylated spirit) placed in a lightweight wooden case. The lamp is clipped to a false sliding floor. Inlets and vents are cowled. Glass or perspex window in door. Anti-tip base, held by wing-nut, for standing on rough ground, removed when carried in grafting established trees.

apparatus. A very simple but efficient lightweight heater can be made from wood and an inexpensive oil lamp as illustrated. (Fig. 20.)

For indoor work any small, stationary oil heater will prove adequate.

WAX BRUSHES
Small paint brushes (sash tools) are excellent for wax or other liquid seal application. Where the wax is applied hot it is necessary to heat the brush in the wax and sooner or later the brush will be scorched as it rests on the bottom of the wax pot. This can be avoided by fixing in the handle of the brush a small nail which can be hooked over the edge of the pot to keep the brush from resting on the bottom.

Inexpensive brushes are easily made by tying sisal cord firmly at the ends of sticks and fixing this by driving a tack through the ties into the stick, fraying and trimming to the desired shape. Small brushes with tufts not more than 12 mm ($\frac{1}{2}$ in.) wide and about 20 mm ($\frac{3}{4}$ in.) beyond the stick are best.

Tying Materials or Ligatures
Almost any tying material may serve to hold stock and scion together until joined, but of the many materials used some have proved more suitable than others.

RAFFIA
Is probably used more than any other. It is strong yet pliable, lasting yet impermanent; moreover it is inexpensive when purchased in quantity. There are various qualities; the better grades are less wasteful of time and should be insisted upon. Some operators wet their raffia, but unless all the wetted raffia is to be used almost immediately it is better merely to damp the hanks by leaving them in a moist atmosphere overnight or by wrapping them in a damp cloth. Before the grafting or budding begins the raffia should be hand-picked, tied in small hanks, and cut to a suitable length.

Plastic imitations of raffia are available. They are consistent in width and texture but, being very smooth and lightweight, they are more difficult to hitch or knot than raffia, particularly in windy weather.

RUBBER BUDDING STRIPS

A medium strip is about 100 mm (4 in.) long, 2·5–4 mm ($\frac{3}{16}$–$\frac{5}{16}$ in.) wide, and capable of being stretched to 0·6 m (2 ft.) or more without breaking. In the absence of budding strips, some English tree-raisers have made use of discarded inner tubes from bicycle wheels. The tube is slit down its length, opened out, and passed through an ordinary office guillotine used for trimming papers, so that the strips are cut with straight sides. Notched strips, resulting from the use of scissors, tend to break when stretched. The strips are fixed by lapping the first turn over the beginning of the strip and tucking the end under the last turn. Advantages claimed are that no constriction occurs in the region of the bud, as so frequently occurs with neglected raffia and the other ties, and no attention is required for retying or removals.

Ordinary rubber bands 0·75 × 5·5 mm ($\frac{1}{32}$ × $\frac{7}{16}$ in.) with a circumference of 200 mm (8 in.) have been cut and substituted for budding strips. They are quite easy to obtain and are suitable for garden use, but where large quantities are required the straight budding strips will normally prove less expensive.

Strips of natural rubber must *not* be smeared with petroleum jelly (*see* page 120) for this causes them to perish and break too soon. Synthetic rubber strips are not affected to the same degree and though they are generally not quite so easy to adjust, being less resilient, they are preferred where petroleum smears are needed.

PLASTIC STRIPS

Many widths and colours are used. If much above 6 mm ($\frac{1}{4}$ in.) wide the strip is not readily hitched or knotted, particularly if it is one of the thicker grades. Strips may be home-made from polyethylene sheet or lay-flat tubing of 50- to 100-gauge (0·002–0·004 in.), but it is usually far better to purchase the strips ready-made. A description of two kinds may suffice. One is a polyvinyl chloride (PVC) strip 5 × 0·25 mm ($\frac{3}{16}$ × $\frac{1}{100}$ in.), ribbed for ease of gripping, and cut in lengths of from 200–380 mm (8–15 in.). This strip will stretch to about twice its normal length, but when applied it does not release itself by breakage and must be removed before constriction occurs. However, there is sufficient stretch to suffice until the graft union is complete, and the scion has made considerable growth. Thus PVC strip has proved very useful for tying both bud grafts and dormant grafts of woody subjects and is admirable for rapidly-growing scions liable to constriction following the use of rigid ligatures such as raffia. It is an advantage to have an orange-coloured strip because this is so clearly seen in the nursery. The non-perishable nature of PVC makes it possible to use each strip many times. To facilitate salvage the strips should be drawn out to their full length, immediately before binding, so that they are long enough to hitch with a loop (Fig. 21, A) for easy removal. PVC is chemically inert and is highly resistant to mineral oils and greases, so is unaffected by petroleum (*see* page 120). When collected for re-use the strips are washed in hot water and domestic detergent to free them of grease and dirt.

A thin plastic strip, particularly useful for bud-grafting, is available in reels. It is about 6×0.075 mm ($\frac{1}{4} \times \frac{3}{1000}$ in.), can be broken into lengths by a quick pull, and being much more delicate than the PVC strip may be left in position to free itself as the graft grows, provided it is not bound layer on layer. In spite of assurance that this smooth plastic is self-adhesive, it should be half-hitched or knotted when used in the open nursery. It is chemically inert and may be obtained opaque or transparent.

Thin plastic tapes in various widths suitable for particular purposes are readily obtainable. Narrow tapes are more readily hitched or tied than wide but require more turns to cover the work. They must be sufficiently strong for firm tying yet readily snapped from a reel or dispenser. Tapes up to 25 mm (1 in.) wide are commonly used, either clear or white.

RUBBER SHEET FOR HERBACEOUS MATERIAL

When grafting thin and delicate stocks, such as petioles and herbaceous sprouts, it becomes difficult to tie the graft with the usual ligatures. In these cases a sheet form of self-sealing, pure crêpe rubber[58] has proved useful. A piece of rubber sheet is folded around the graft and pressed together so that it holds and seals the graft (*see* Figs. 26, 68 and 69). This is then covered with a sheet of tinfoil to prevent deterioration of the rubber in sunlight (*see* page 135).

RUBBER PATCHES

Rubber patches[34] up to 40 mm ($1\frac{1}{2}$ in.) square, fixed on the opposite side of the stock by means of a wire staple, have proved a rapid means of tying inserted buds where no very high pressure is required (Fig. 21). One 'tyer' can keep up with two budders.

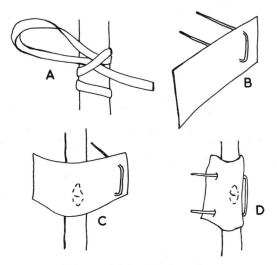

Fig. 21 Tying bud grafts

A. Plastic (PVC) strip looped to facilitate removal. B. Rubber sheet and steel-wire staple[34]. C. Sheet placed over the bud and pulled firmly round behind the stock. D. Staple pushed through and held by tensed sheet.

WAXED COTTON OR TWINE

Waxed ties last longer and are self-fixing. They are commonly used for tying bench grafts which are afterwards placed in moist situations, requiring no sealing against drying. Soft light twine or cotton (crochet cotton) are commonly used. The spools are dipped, after locating the beginning of the twine, into melted wax for up to half an hour and allowed to drain. When grafting, the spool may be kept soft by being placed in warm water. It is unnecessary to knot the tie, it is merely broken off and remains in position. Where many thousands of bench grafts of one kind are made, as in America, tying machines are commonly used for binding the grafts.

VARIOUS FIBRES

Many manufactured threads, oiled wool and cotton, and innumerable naturally-occurring fibres may be used with success. Provided a material is non-poisonous to the stock and scion, is pliable, does not shrink, and is easily manipulated, there is no reason why it should not prove suitable as a graft ligature.

Other Fastenings and Wrappings

NAILS

Sometimes scions are nailed to the stock and not tied or bandaged. The most suitable nails are thin in the shaft and have comparatively large flat heads. Gimp pins are eminently suitable. An appropriate gauge should be selected: 9 mm ($\frac{5}{8}$ in.), 20-gauge, are most commonly used, but longer pins may be necessary for thick scions. When grafting herbaceous plants or succulents it has been found convenient to use a natural nail, namely a thorn or spine, to fix the parts together.

GRAFT CLIPS AND METAL STRIPS

Various clips, pins and pieces of malleable metal have been used to provide a temporary anchorage for grafts, particularly where protected against mechanical disturbance. In the earlier work with cucurbita grafts in Holland a modified paper clip, of rust-proof wire, was used (Fig. 77). This was superseded by lead or foil strips which were simply wound round the graft (Fig. 27). When wedge grafting herbaceous stems foil strips may be wound first round a pencil, then slid over the prepared stock ready to be moved up when the scion is in place.

METAL FOIL WRAPS

Aluminium foil has been used to wrap cleft grafts. Pressed in at top and bottom to form a moist chamber, it permits exuded sap to drain away. The foil reflects sun heat, reducing the risk of overheating. Foil wraps are used in combination with plastic bags which are drawn over the wrapped graft. The bag is holed or slit in one corner for aeration and exit of the new growth[126]. Foil is also used to delay deterioration of crêpe rubber seals (Fig. 26).

BAGS, SLEEVES AND COVERS

Where plants are grafted *in situ* in the open, or bench grafts are set out at an early stage, some temporary protection may be helpful in raising the percentage success and may also speed development of the plant. Individual protectors may be made from waxed paper, or from bitumen-lined paper such as 'Sisal Kraft', folded into either an open-ended cylinder or a cone fixed by a pin or staple. Cylinders provide adequate ventilation, avoiding heat build-up, but cones should remain open at the apex. A little soil heaped around the base of the protector prevents it blowing against the graft. If the grafted plant is staked the protector can be passed over the stake and plant. Polyethylene sleeves and bags may be used but they must be well ventilated, preferably by having a number of holes in their sides. Merely leaving the top open will not serve, because the soft material may fold over and adhere to itself.

Thin plastic sleeves may be drawn over apical grafts (Fig. 70) to prevent wilting of delicate scions and to give ready access for inspection and gradual ventilation.

Containerized plants may be placed in plastic bags for temporary protection following grafting. The bag is holed for drainage and the mouth is closed over the scion by a clip, rubber band or bow tie. A graft may also be protected by a lamp glass (Fig. 71).

SPLINTS

Most well-designed grafts splint themselves either by over-lapping of the components or by firm wedging, but some herbaceous grafts and all plain end-to-end (abut) grafts must be supported until the union becomes strong. A simple method with hollow-stemmed plants is to insert a rod of wood or other material half into the stock and half into the scion. In solid stems, particularly those of cacti, one or more toothpicks or similar articles are pushed into each component until the cut surfaces are in firm contact. External splints are necessary when double-working with intermediates in the form of transverse slices in experimental work (Fig. 77).

Thin, delicate scions and stocks may be held together by pushing them into either end of a glass tube. Simple transverse cut surfaces may be brought in contact in this way but it appears rather better to make slanting cuts so that the parts overlap. Tubes with bores of from 4 to 7 mm ($\frac{5}{32}$ to $\frac{9}{32}$ in.) have been used. Very thin glass, which breaks as the shoot grows, is necessary to avoid damaging the new union[107]. Such thin tubing is made by first blowing a glass bubble and then drawing it out.

Bench-grafting techniques for splinting and anchoring of very small, delicate components, using short sections of plastic tubing and slivers of bamboo, or toothpicks, have proved highly successful (Fig. 95).

SHOOT GUIDES
These are metal clips placed on dormant budded rootstocks to compel straight growth from the bud. (Fig. 22) They are not at all essential but sometimes prove useful where extra-straight growth is required at the junction of stock and scion.

Fig. 22 Shoot guide clips
Pliable metal clips which have been used to obtain very straight growth from buds. Budded rootstocks are cut down to the bud before applying the clip in the spring.

CHAPTER VI

Methods of Grafting

Seeing that almost any carpentry which achieves cambial contact between scion and stock may lead to a successful graft, it follows that the forms of grafting may be very numerous. Many hundreds of both simple and intricate techniques have been described in the literature, and the possible variations on what may be called the basic methods are infinite. It is small wonder that writers hesitate to classify the existing grafting methods, yet an arrangement under headings is an aid to the study of the art. All grafting naturally falls under two heads: (1) grafting by approach, in which scion and stock are not, or only in part, severed from the parent plant until a union is effected, and (2) grafting with detached scions. Approach grafting is necessarily somewhat cumbersome, and possibly for this reason the technique has not been developed in very many directions. Grafting with detached scions is practised in great variety and it will only be possible to describe here the more practical and interesting methods.

THE MAIN DIVISIONS OF GRAFTING

1. **Approach Grafting:**
    a. True approach grafting.
    b. Inarching.
    c. Bridging.

2. **Detached Scion Grafting:**
    a. Bud grafting (budding).
    b. Inlay (veneer) grafting.

    c.   Apical grafting.
    d.   Side grafting.
    e.   Bench grafting (including cutting grafting).

Rind Grafts and Cleft Grafts

Any method involving the separation of the rind from the wood is, to speak truly, a rind graft. Its use is restricted to seasons of active growth of the cambium, whereas cleft grafts, which do not depend upon the separation of the rind, may be employed at almost any season. Thus all methods which are not rind grafts are cleft grafts, including the sliced and inlaid types such as the whip-and-tongue and the kerf. So it is that each division contains its cleft and its rind methods, as will be seen.

AUTUMN GRAFTING OUTDOORS

The popular times for outdoor grafting are late winter and early spring with leafless, 'dormant' scions, and late summer with defoliated shoots. Other times for outdoor work are late spring with overwintered scions and autumn with mature shoots when their leaves are falling. Autumn grafting is highly successful with many woody plants; there is still sufficient warmth to encourage healing, and components share the winter chilling period so that they become active together in the spring. Plants which are prone to 'bleed' at other times, notably maples, mulberries and walnuts, may be readily grafted in the autumn, either on the bench or in the open, provided they are protected from winter killing frosts. High-worked standard trees, including *Malus* and *Prunus* spp., generally make better yearling heads when autumn grafted; whereas spring grafts are adversely affected by sudden temperature changes in spring resulting in poorer growth. Open ground autumn grafts should be extra carefully sealed.

i. APPROACH GRAFTING

a. True Approach Grafting

The distinguishing feature of approach grafting proper is that the plants to be joined are brought together whilst each retains parts above and below the point of contact. Such true approach methods are here divided according to the method of carpentry. Inarching differs from approach grafting in having no part of the stock above the grafting position.

Approach grafting has its uses in joining plants which otherwise unite only with difficulty. Both scion and stock are sustained by their respective parents until a union is formed, and provided the parts are placed firmly together and are mutually compatible, they grow together sooner or later. Approach grafting can be successfully accomplished at any time of the year though, naturally, rind methods can only be used when the tree is in active growth. Callussing will take place slowly in cold weather, but, provided the work is well done and sealed, a good union will form in the spring.

When a sound union has formed, the stock above and the scion below are severed so that the scion and stock are solely dependent upon each other. This cutting away of scion-base and stock-apex may be done at one time but more often the stems are gradually severed so that the operation is spread over a number of weeks.

SPLICED APPROACH GRAFT

This is the simplest graft of all and is merely an aided natural graft. The stock and scion, preferably of equal size, though this is not important, are each sliced (Fig. 23, A) to expose the cambium in, as far as possible, equal pattern. These cut surfaces are placed together and tied securely. Herbaceous plants, under glass, require no sealing but otherwise the graft should be sealed.

The spliced approach graft can be modified to meet special needs. In India it has been used for many years in the propagation of the mango. Where the only available scions are considerably larger than the stocks, two stocks may be set against one scion (Fig. 23, B). It has been noted that young seedlings approach grafted to mature trees have flowered below the union, but only in those cases where the mature shoot was ringed (Fig. 24) below the graft.

In so-called embryo grafting, the spliced approach has been used to join young seedlings. When the first true leaves appear, one of the cotyledons is removed by a slicing cut

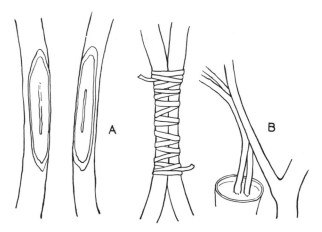

Fig. 23 Spliced approach graft
A. The components sliced to expose the cambium in equal pattern. B. Two small stocks set against one scion.

Fig. 24 Seedling approach-grafted to a ringed mature shoot
A. Mature shoot ringed. B. Seedling in pot, fixed level for watering. C. Shoot and seedling prepared. D. Held and sealed by adhesive tape.

beginning just above the cotyledon and ending about 25 mm (1 in.) lower. The partner plant is treated similarly and both are bound together and replanted. When a union is formed, the top of the stock is removed and later, when transplanting, the scion-root is cut away.

TONGUED APPROACH GRAFT

This has also been named the English method[3]. Comparatively long, deep slices of rind and wood, usually about six times as long as the diameter of the smaller component, are removed from both scion and stock so that the cambia are equally exposed. A tongue, half the length and in the middle of the sliced portion, is cut downwards on the stock and upwards on the scion (Fig. 25, A.). Stock and scion are brought together (B), tied firmly, and sealed. Herbaceous subjects and plants under glass may often be left unsealed. When union is complete, the scion should be severed from below and the head of the stock removed from above the union, preferably in three or four stages.

This method is used when investigating the transmission of virus in various plants and has been adapted for work with the strawberry (*Fragaria* spp.)[58]. The stolons are joined most successfully whilst young and when only the first leaf is

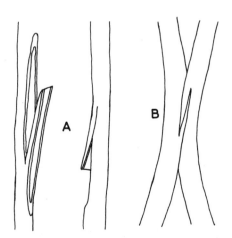

Fig. 25 Tongued approach graft

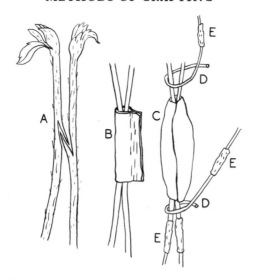

Fig. 26 Approach grafting the strawberry

A. The young stolons tongued, without preliminary slicing. B. Fitted and wrapped with sheet of self-sealing crêpe rubber. C. The rubber covered with metal foil to exclude light. D. Lead wire supports. E. Bands of grease to prevent migration of insects.

unfolding from the new runner. The initial slicing, described above, is dispensed with and each tongue is made by a single cut (Fig. 26, A). A suitable knife for this delicate operation is the Bard Parker knife described in Chapter V. The tongues are interlocked (B) and the graft bound with a small piece of proprietary sheet-form, self-sealing, pure crêpe rubber. This rubber is then covered with metal foil to exclude light, which otherwise causes the rubber to perish. The completed grafts are tied up to supports, using lead wire (22-gauge). All lateral branches of the stolons are removed as they appear. The stolons may be grease-banded to prevent the migration of mites and other crawling insects. Young leaf-petioles have also been grafted successfully by this method.

A simplified form of tongued approach graft is used on a commercial scale in Holland to combat fusarium wilt of cucumber. The disease-susceptible edible cucumber (*Cucumis sativus*) is united with the highly-resistant *C. ficifolia* by approach grafting in the seedling stage. Seeds of edible

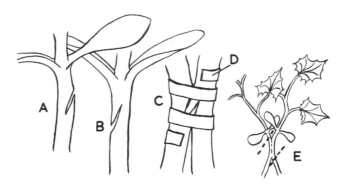

Fig. 27 Approach grafting the cucumber

cucumber are sown in a high temperature (25–30°C, 77–86°F), followed one week later by sowings of the faster growing *C. ficifolia*, so that the young plants develop long hypocotyls for grafting. Nine days after the second sowing the plants are lifted and grafted (Fig. 27).

A razor cut is made below the cotyledons of each, downwards in the *C. ficifolia* rootstock (A) and upwards in the *C. sativas* scion (B), each cut penetrating halfway through the hypocotyl. The plants are interlocked (C) and secured by two or three turns of metal foil or lead strip (D). The grafted plants are placed in one pot and, after ten days at 25°C (77°F) the *C. sativus* is removed from below, and the *C. ficifolia* from above, the union (E). After another ten days or so they are set in their permanent positions.

This simplified tongued approach method is also used in Holland for tomato where a high water table prevents the full effect of soil sterilization and it becomes necessary to employ disease-resistant rootstocks (*see* page 236). Seed of a vigorous, disease-resistant selection is sown up to fourteen days before that of the scion cultivars to obtain matching development for grafting. This interval becomes less as the season advances; by March it may be only three or four days. When the rootstock plants are 125 mm (5 in.) high, and the scion plants 75–100 mm (3–4 in.), the plants are lifted; a single cut is made downwards in the stock and upwards in the scion, the plants are interlocked, taped or clipped together, and placed in a sterilized medium. After some fourteen days they are lifted, the scion root and rootstock top are cut away and the composite plant is set out. If only 'corky' infection is feared, both root systems are left intact to provide additional anchorage at this critical period.

INLAY APPROACH GRAFT

When one of the partners is very large it is impracticable to expose the cambium by slicing, and the inlay or slot method affords a useful alternative. The small component is prepared by a long slicing cut (Fig. 28, A) followed by very shallow straightening cuts at the sides (B). The large component receives two parallel cuts (C) through the rind, spaced the width of the smaller. The rind is removed from between these two incisions and the parts fitted together and fixed by two thin nails (D) or gimp pins. Sealing completes the operation.

If the rind does not lift readily, the smaller component may be prepared by two cuts (Fig. 28, E and F), forming a long wedge, and the larger component grooved by two deep

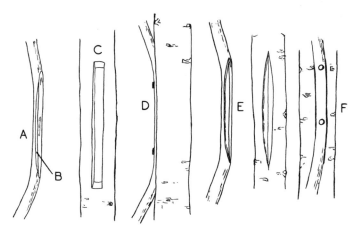

Fig. 28 Inlay approach graft

A. The smaller component prepared by one main slicing cut and two very small straightening cuts (B). C. The rind of the larger component removed from between two parallel incisions. D. Fitted and held by two small nails. E. Method used when rind does not separate readily from the wood. F. Laid in and nailed.

incisions, so that when fitted the cambia of the two partners are tightly pressed together. Nails are used to fix the graft and sealing is helpful.

RIND APPROACH GRAFT
Where the partners are of very unequal size the smaller is prepared by a single cut (Fig. 29, A) and is placed under the lifted rind (B) of the larger. For firm fixing two fillets of wood (C) are placed on either side of the inserted member, and the string or other tying material passes over these so that the pressure is greatest where most needed. All cut surfaces should be sealed.

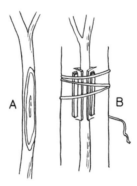

Fig. 29 Rind approach graft
A. The smaller partner sliced. B. The smaller placed under the lifted rind of the larger partner. Two fillets of wood under the tie serve to apply adequate pressure where it is needed.

BOTTLE GRAFTING
A method of approach grafting commonly termed bottle grafting is sometimes employed when one of the components, the scion, is a small branch severed from its parent. This rootless branch stands in a bottle of water until a union is formed (Fig. 30). A few lumps of charcoal placed in the water serve to keep it sweet and clean. At no time must the water fall below the level of the cut end of the branch. Leafy scions are best collected from the parent by severing under water, the cut end remaining under water until a union is assured.

With very readily-grafted subjects the bottle may be

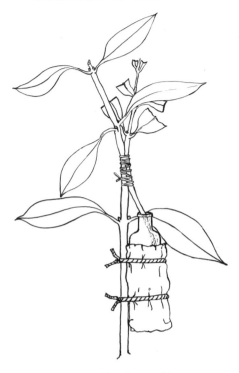

Fig. 30 Bottle grafting
The leafy scion is approach-grafted to the established stock. The base of the scion is kept in water until a union has formed, then this base is cut off close to the stock. The head of the stock is also removed, preferably by stages.

replaced by a very small lightweight plastic bag containing a small amount of water or nutrient liquid. This technique serves to separate the root systems of the components, thus preventing contact with soil-borne disease which occurs when the scion roots are allowed to enter the soil.

APPLIED GRAFT
Two plants of equal size may be split longitudinally and half of one be joined to a half of the other to form a composite plant. Such grafts are often successful but attempts to join the existing apical meristems into a single growth have not succeeded.

Applied grafting has been used to investigate rootstock

effect upon scion varieties of apple. Where two half rootstocks were applied and used as a rootstock for a single scion, each half rootstock conveyed its effect to the part of the scion directly above, though the main factor in determining the relative vigour of scion branches was the orientation of the topmost branch, the orientation of the stocks relative to the branches being of secondary importance.

b. Inarching

Under this heading come all those methods in which one of the partners is decapitated and inserted into the other so that there are two parts below the graft and one part above.

CLEFT INARCHING

When neither of the components is more than 50 mm (2 in.) in diameter a simple incision in one stem (Fig. 31, A) serves to receive the prepared end (B) of the other. The angle of approach depends upon the angle of the incision. A short, deep incision is needed where it is desired to transfer the

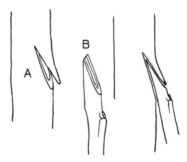

Fig. 31 Cleft inarching

A. An upward cut forms a cleft. B. A wedge is made to fit into the cleft. This graft must be firmly tied.

shoot to the inserted stock as rapidly as possible, but where the inarching is designed to add to the vigour of the incised component, only a shallow cut is required. This graft should be well bound and sealed. Waxed cloth or tape is ideal for this purpose.

RIND INARCHING

Where one component is much larger than the other it is convenient to insert the smaller under the rind of the larger, provided the rind parts readily from the wood. There are many ways of doing this. One of the more practical methods is similar to the veneer crown graft, described on page 258, but inverted. The thinner component is prepared by one principal cut (Fig. 32, A), two very shallow cuts (B), and the extreme tip is removed (C) to strengthen the end and expose the cambium. Two parallel incisions (D) are made in the rind of the larger component and the lower ends of these incisions are joined by a third incision (E) inclined upwards at 45° through the rind. It is now possible to raise the rind slightly and to insert the thinner component. The rind is pushed outwards and some of it may be cut off. Two nails serve to hold the parts firmly and sealing completes the operation.

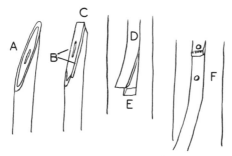

Fig. 32 Rind inarching

A. The smaller partner receives one long slicing cut. B. Two very shallow side cuts are made merely to straighten the edges. C. The thin tip is removed by a cut on the opposite side to the first. D. The larger partner receives two parallel incisions. E. An inclined incision permits the raising of the rind. F. The components fitted, nailed, and surplus rind removed.

Another method is to insert into an L-shaped incision. The part to be inserted is prepared by making one principal cut (Fig. 33, A) and a second cut (B) on the opposite side to expose the cambium. This technique, except for the inversion, is as described for the inverted L rind graft on page 269 and, as described there, a thin shaving may be taken from the edge of the principal cut (A). One vertical radial cut (C) is

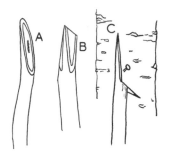

Fig. 33 Rind inarching with an L incision

A. The principal cut upon the member to be inserted. B. A second shallow cut to expose the cambium on the opposite side, and a thin shaving to straighten the edge. C. A vertical incision in the rind, joined by an upward inclined cut, permits the rind to be raised. The prepared end is pushed under the rind and fixed by nailing.

made in the rind of the larger partner and an upward inclined incision connects this in the form of an open L. The prepared end is inserted under this rind and fixed by a thin nail. The work should be well sealed.

c. **Bridging**

The chief use of this method is for the repair of trees as described in Chapter VIII, but it is also of use in refurnishing bare stems and branches with young growth, in frameworking and for many other purposes (Plate 27).

The scions can be of almost any length, but short scions are difficult to manipulate. The ends of the scion are inserted into the stock in any convenient manner. The L and inverted L rind graft, the side cleft, and various inlay methods are all suitable. With very long scions it is possible to make two or more spans by inlaying the scion at intervals. Fixing may be done by small nails or gimp pins, and sealing completes the work.

Bridging proves useful in covering bare trunks and branches which might otherwise suffer from sunscald in hot weather.

In some cases it is possible to strengthen or support weak branches by connecting them by a 'bridge' from a stouter limb. A suitably-placed lateral shoot on the stout limb may be inarched, or approach grafted, into the weaker, thus merely

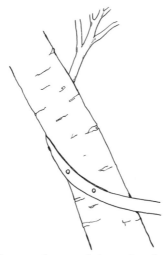

Fig. 34 Supporting weak branches by bridging

A lateral shoot from a strong limb is inlayed around the branch requiring support. The 'bridge' enfolds the branch in a firm anchorage whilst the junction is established.

involving one operation. In such strutting, the part grafted in must withstand the movement of large limbs, and strong junctions are vital. Such can be assured by inlaying the thinner partner partly around the thicker (Fig. 34). Even so, it is almost always necessary to support the parts until a really strong junction has formed. Where large, heavy limbs are already tending to split apart, they may be temporarily braced by a Spanish windlass of stout rope or hose-covered woven wire.

2. DETACHED SCION GRAFTING

a. Bud Grafting

Grafting with a single eye, or bud, is often termed budding and takes many forms. The most popular is that known as shield budding, so named because of the appearance of the prepared scion. This method, now to be described in detail, is simple and highly efficient. There are more complicated bud grafts and some of these will also be described, since each has been designed to overcome particular difficulties.

SHIELD BUDDING

This is most easily employed where the portion of the stock to be budded is not more than two or three years old, or where the rind is not so thick as to cause manipulative difficulties. It is normally carried out during the height of the growing season when the rind parts readily from the wood. Special budding knives, as described in the previous chapter, are necessary.

The scion-buds (eyes) are usually taken from shoots of the current season's growth where they exist in the axils of the leaves. For early summer or springtime budding the previous season's shoots provide the buds. The collection of the scion-wood has already been described in Chapter IV and a fuller description of budding, as practised in fruit tree nurseries, appears in Chapter VII.

The bud stick is held by the upper end and the first good bud nearest this is removed by a shallow slicing cut which begins half to one inch below the bud (except where the inverted T method is employed as described below), passing under the bud and coming out well above it as shown in Fig. 35 (A). If the knife blade be arrested before it quite attains the

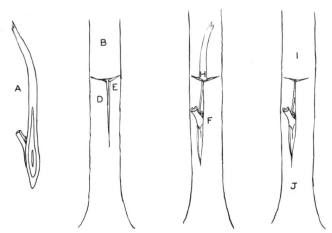

Fig. 35 Shield budding

The bud (eye) prepared and inserted in the stock ready for tying. Refer to text for full description.

surface at the finish of the cut, and the removal of the bud patch is completed by tearing, the patch, or shield, will terminate in a long strip of rind which makes a convenient handle. This handle should be cut off after the insertion of the bud (H). The stock (B) is prepared by making a T-shaped incision (D) through the rind down to the wood. The rind is raised (E) and the bud inserted (F) by sliding the shield downwards under the lifted rind so that the bud lies between the edges of the rind and well below the horizontal incision. The tail or handle (H) is now cut off exactly at the horizontal incision and the bud is firmly tied.

Many tying materials have been used with success. Raffia is still commonly used, but rubber budding strips are now employed on a considerable scale. Home-made and proprietary adhesive tapes are also used. Whatever the material the bud must be firmly tied in position for two or three weeks, and thereafter the stock must not be constricted. The use of various tying materials is described in Chapter V.

Modifications of shield budding. Almost every separate movement in the act of budding has been modified to suit the operator's particular convenience or, less commonly, because such variations in technique have proved more efficient.

Removal of wood from the shield: Very many operators insist upon removing the small sliver of wood within the bud

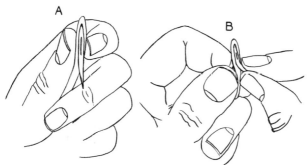

Fig. 36 Removing the wood from the bud shield

A. The sliver of wood has begun to separate from the rind and the rind 'tail' is gripped to the base of the thumb by the second finger. B. The wood is held between finger and thumb, of the other hand, and flicked upwards from the shield.

shield, though budding with the wood left in the shield rarely reduces the chances of success[37]. In the method described above the shield is cut thin, but where the wood is to be removed a thicker shield may be used. To remove the wood, bend the tail (Fig. 35, H) to and fro so that the layer of wood separates from the thin rind in the tail. Now place the petiole between the finger and thumb of the left hand, with the cut surface uppermost, so that the tail (Fig. 36) lies along, and in line with, the thumb. The second finger should now grip this tail against the base of the thumb (Fig. 36, A), and the wood may be flicked upwards and backwards, increasing the bending motion slightly (B) as it is about to leave the actual bud or eye.

The little knob or heart-shaped protuberance (bud-trace) should remain attached to the shield. It is not always vital that this should stay but it is generally considered that a hollow shield should not be used and such are usually discarded. However, with apples and pears at least, the presence of an intact bud-trace is immaterial.

Another way of obtaining woodless buds is to leave the wood attached to the bud stick. The stick is held with the upper end towards the operator and the knife entering below the bud, as already described, is withdrawn after passing no more than 12 mm ($\frac{1}{2}$ in.) beyond the bud. A straight cut is then made, through the rind only, above the bud, to connect with the first cut at either side. The base of the petiole is firmly grasped with finger and thumb and the shield is lifted off the wood by a sliding-with-twisting and lifting motion and immediately inserted into the stock.

The inverted T incision. The use of an inverted T incision (Fig. 37, B) in the rind of the rootstock is sometimes advocated for moist conditions because water does not so readily enter the incision and cause rotting. The inverted method is traditional in the budding of citrus[25]. This inversion entails a different preparation of the bud shield, which must now be held with the base towards the operator and the knife proceeds from half an inch or so above the bud to a point below the bud. The wood may or may not be removed, as already described, and the bud shield is inserted

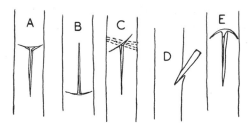

Fig. 37 Incisions used when shield budding

A. The normal T incision. B. The inverted T incision. C. Slanting incision crossed by the tie. D and E. Inclined incision which closes by lapping after complete entry of the tail-less budshield.

and drawn upwards beneath the lifted rind of the stock. Except for this, the technique is the same as for normal shield budding.

Slanting incisions. The incision which crosses the vertical incision may be inclined at an angle of some 45° with the vertical so that the tying material crosses this incision rather than runs parallel with it (Fig. 37, C). This effect is accentuated if the inclination runs opposite to the spiral path of the tying material.

Inclined incisions. Another variation is to incline the knife at 45° with the vertical axis of the rootstock so that the cross incision is closed by lapping (Fig. 37, D and E) after entry of the bud. The vertical cut is made first. As the second rolling cut is completed the knife is rocked outwards, lifting the rind flaps, facilitating entry of the prepared bud. Rind flaps prevent the use of tailed buds (Fig. 35, A), and thus the bud must be completely prepared before insertion so that it can be pushed down to lie flat on the cambium of the stock.

Incisions in the form of a cross. This method has been suggested for the insertion of large buds into small stocks. The upper end of the shield is slipped under the rind of the stock above the horizontal incision so that the shield is firmly held at both ends. As the same purpose is served by lengthening the vertical incision, in the normal T method, the cross incision is hardly necessary.

Methods of tying: When using raffia, soft string or rubber strips, one of the commonest methods is to begin tying with a half-hitch (Fig. 38) below the lower end of the perpendicular incision and to bind upwards, taking special care to bind firmly immediately below and above the bud. The bud must not be caught under the tie and crushed, nor constricted at its base. The tie is finished off with two half-hitches somewhat above the horizontal incision in the stock. Loose ends of raffia or string ties are best cut off at this time.

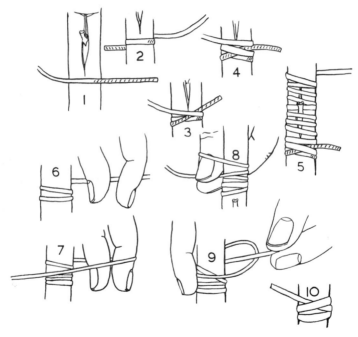

Fig. 38 A method of tying inserted buds

The tie is laid across below the vertical incision (1) and passed behind the stock (2). The short end is held by the next turn in front (3) and a further turn catches the short end down against the previous turn (4). Binding upwards, carefully but closely passing the bud (5), the tie is finished off well above the incisions. In this the tie is held by the right hand, the finger and thumb of the left are placed within the encircling tie, against the stock (6), and draw the tie back behind the stock (7) where they take it from the right hand (8). The first finger of the right hand is now pressed against binding, to prevent loosening, and the tie is drawn through (9) and pulled around the stock to form a half-hitch (10). A second similar half-hitch is added to make a sound job.

With prominent or swollen buds, often found in pears and cherries, there is a danger of displacing the bud by tying from below upwards. In such cases it is better to begin with a half-hitch close above the bud, then to proceed downwards before coming upwards in an open spiral to finish above the horizontal incision. An alternative method, not possible with rubber strip, is to place the middle of the length of tying material either above the horizontal incision or just above the bud and to cross the two ends as in figure-of-eight bandaging, finishing with a reef knot, usually behind the stock. Many find the knotting method more difficult than half-hitching a single strand.

With suitable adhesive tapes it is possible to tie efficiently with one and a half spiral turns. Tapes vary in width but a 25 mm (1 in.) tape is suitable for most purposes. These tapes should not be lapped layer upon layer, not only because this is wasteful but also because the growth is thereby constricted and the tape is difficult to remove.

Use of dormant buds ('June' budding). Known in U.S.A. as 'June' budding because it is done earlier than normal shield budding. As soon as the stock rind can be lifted, but before the current season's scion buds are available, dormant buds are taken either from stored scion-wood or from dormant buds at the base of shoots, and are inserted in the normal manner. When successfully united the buds may be forced into growth by removing the stock above them or by bending, as described later in discussing the aftercare of buddings.

DOUBLE SHIELD BUDDING

This is used to overcome stock/scion incompatibility of the kind resulting in breakage at the union, e.g. between certain pear scions and quince rootstocks. There are two rather similar methods, the first is fully described in Chapter VII (Fig. 99), the second, known as the Nicolining method, after its inventor and patentee[99], is described here. Two yearling bud sticks are required, one of the variety to form the upper part of the tree and the other of a kind known to be compatible with both the rootstock and upper scion. The risk of interchange of the bud-sticks is reduced by clearing the intermediate variety of both leaves and petioles. A normal T-cut is made in the rootstock. A bud is sliced from the intermediate and discarded. A second slice (Fig. 39 A) about 1·5 mm ($\frac{1}{16}$ in.) thick is collected. A normal bud (C) is cut from the upper scion and fitted to the slice (B) and together inserted into the T-cut as shown (D). Binding completes the operation. Early budding with ripe wood is advocated and considerable care should be taken to match the cambia of slice and bud shield.

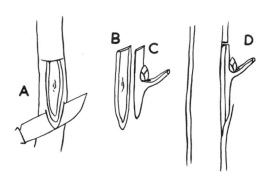

Fig. 39 Nicolin's double-shield budding

PATCH BUDDING

Under this heading come all methods of bud grafting, with a single eye or bud, in which the rind of the scion replaces a part of the rind of the rootstock. The size and shape of the piece may vary but the method is still patch budding. The patch is more difficult than the shield method and is only employed where it has proved to give better results, as in the walnut and pecan, the rubber tree (*Hevea brasiliensis*), and many other tropical subjects.

For ease of manipulation the stock and scion should be of approximately the same age and size and in such condition that the rind parts readily from the wood. When the rind of the stock is considerably thicker than the rind of the scion, the stock rind should be pared down so that, when in position, the scion is either level or above the stock rind. This facilitates firm tying of the patch[111].

By the use of a knife having two parallel blades, as

described on page 108, cuts (Fig. 40, A) are made horizontally round the bud wood, through the rind above and below the selected bud. Two vertical cuts (B) made with a single blade connect the horizontal cuts, and the patch (D) containing the bud is removed. The stock (E) receives similar horizontal cuts (F) slightly longer than the width of the patch. A single vertical cut (G) enables the rind to be raised and peeled back (H) so that the bud may be placed in position. The stock rind is peeled back a distance equal to the width of the bud. A vertical cut (J) removes the surplus stock rind, and all exposed edges of stock rind and scion patch are in contact. Tying completes the operation. In some cases it appears to be an advantage to seal the exposed cut surfaces.

Fig. 40 Patch budding

FLUTE BUDDING

Sometimes known as modified patch budding, this method is used when it is required almost completely to interrupt the connection between the upper and lower parts of the stock at the graft.

The scion is prepared by making two horizontal cuts (Fig. 41, A) encircling the bud-wood (B), through the rind down to the wood. This operation is facilitated by the use of a knife with two parallel blades. A vertical cut (D) connects the two horizontal cuts and enables the bud to be removed. If the horizontal length of the bud patch is more than sufficient to encircle some seven-eighths of the stock, it should be shortened by a vertical cut removing the required amount of surplus rind. This will avoid completely girdling the stock when fitting the bud patch, and, if the bud patch dies, the head of the stock will be saved. Similar horizontal cuts (E) are made almost completely round the stock and a vertical cut (F) enables the rind to be raised from the wood. The vertical edge (K) of the rind patch carrying the bud is placed against the corresponding cut edge (F) of the stock. The rind of the stock between the horizontal cuts is then peeled back until the bud patch fits the stock. The raised rind of the stock is removed by a vertical cut to coincide with the edge of the bud patch. Opposite sides of a budded stock are shown at (H) and (I). Tying completes the operation.

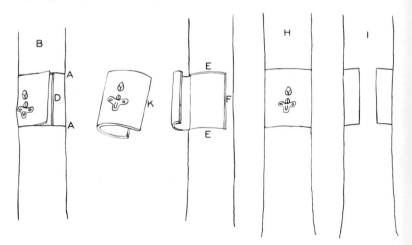

Fig. 41 Flute budding

RING BUDDING (*annular budding*)
This method is an extension of the flute method so that the stock is completely girdled by the scion and beyond this needs no separate description. Should a ring bud fail to unite with the stock the parts of the stock above the ring eventually die.

TUBULAR BUDDING

A flute method of limited use, which might be termed 'tubular' budding, deserves mention. The scion must have the same diameter of wood cylinder as the stock. The stock is cut through transversely and a tube, or ring of rind, about 25 mm (1 in.) long, is removed from the end. A tube of rind with bud attached is slid off the end of the scion-wood and over the end of the stock. Waxed cloth serves to fix the scion and to seal the end of the stock.

FORKERT BUDDING (*brown budding*)

Since its development in Indonesia, primarily for rubber (*Hevea brasiliensis*), this method has proved excellent for many widely-differing species. Where, as in rubber, the subject exudes latex or resin when cut or bent, the method provides means of reducing the risk of exudates contaminating the grafting surfaces. This is achieved (see below) by keeping the rind of the bud shield unbent whilst removing the sliver of wood. The procedure is fully described in *Planters' Bulletin* No. 20 (1955), issued by the Rubber Research Institute of Malaysia.

Well-developed bud-wood, brown in colour and leafless, is best collected on the day of budding. It is kept cool and protected from bruising by placing it in a container or soft wrapping. On reaching the nursery the bud slips (patches) are sliced from the bud-wood by means of a sharp kitchen knife cloth-wrapped at its tip for use with both hands. Only a few slips should be cut at one time, to minimize drying.

Stocks, preferably at least 25 mm (1 in.) thick at the budding position, and in active growth, should be wiped clean with a coarse cloth dipped in a fungicide dispersion (*see* page 103) and allowed to dry.

When bud-grafting rubber it is recommended that three cuts should be made through the stock rind, one horizontal, of 20 mm ($\frac{3}{4}$ in.) and from the ends of this two vertical ones 75–100 mm (3–4 in.) long, preferably on the shady side. A dozen or so stocks should be cut in advance of budding to give time for coagulation of the latex. Remove the sliver of wood from the bud slip, taking care not to bend the rind. This can be done by holding the slip tautly by each end and then grasping the sliver of wood with the teeth and bending it away, taking care not to touch or soil the cambial surface of the bud patch. If the bud-trace comes away with the wood the patch is best discarded. The de-wooded patch is placed on a clean board, or work-box, cambium uppermost, and the edges trimmed to fit within the incised patch on the stock, leaving a gap or frame of about 3 mm ($\frac{1}{8}$ in.). The stock flap is now lifted with the spatula end of the budding knife and the bud inserted without rubbing or sliding the cambial surfaces together. The flap is then closed over the bud and tied with either a bandage or palm frond slip and twine. To shade the bud from sunlight, a few leaves are tied above the binding so that they hang over the bud[28].

The details given here relate to an upward-pointing flap but the technique is equally successful using a downward flap.

After about three weeks the ties should be removed and the flaps cut away at the ridge formed near the attached end. In exposed situations the leaf shade should be renewed. Two or three weeks later the bud may be forced into growth by gradually reducing the top of the stock or by complete removal and sealing of the wound.

Modified Forkert budding. Working with plants other than rubber some workers make only the horizontal cut (Fig. 42,

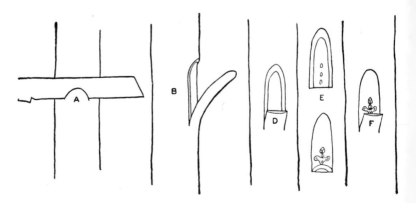

Fig. 42 Modified Forkert budding

A) and then peel the rind (B) to expose the cambium. The rind may peel in one or several strips. Two-thirds of the length of the flaps thus formed are removed (D). The bud (E) is cut to fit the stripped area and, after de-wooding, is placed partly beneath the remaining portion of the flap (F), taped and shaded. (Compare with chip budding, Fig. 45.)

Green strip budding. Following the wide success of the Forkert method, H. R. Hurov of the Department of Agriculture, North Borneo, developed a practical way of bud-grafting rubber when in the young green-wood stage using buds from a new flush on rootstocks $2\frac{1}{2}$ to 5 months old. It is essential that the buds should strip cleanly from the scion-wood. Whilst this technique was specially developed for the somewhat difficult to bud latex producer, *Hevea brasiliensis,* there is no reason why it should not prove successful with a wide range of tropical trees. Working with rubber, Hurov and Chong issued a detailed description[69] of the technique which was found successful and might well be adapted for other subjects and to particular local conditions.

Budwood sources are best established specially to provide successive flushes of green terminal shoots with semi-hardened but flexible stems, though vigorous growth flushes on mature trees are also suitable. Such shoots may be collected by grasping them by their leafy tips and pulling them off, so avoiding bruising the lower parts from which leafless bud patches are later to be stripped. If the rootstocks are established near the source of scions, losses from bruising whilst in transit will be minimized. The leafy tops are cut off immediately after this shoot plucking. With rubber it is necessary to dip the cut ends about 6 mm ($\frac{1}{4}$ in.) into molten wax to prevent contamination of the surface of the wood with latex. The de-leafed bud sticks should be placed in small ventilated bags of thin polyethylene inside containers, or other protective wrapping, and kept shaded and cool until used.

Suitable rootstocks for budding should be vigorous, have ripened (brown) basal bark and be from pencil size up to 25 mm (1 in.) thick. Two 75 mm (3 in.) vertical incisions are made about 8 mm ($\frac{1}{3}$ in.) apart a short way above ground level and are connected at their lower ends by a horizontal rolling incision. The released strip is pulled gently upwards and cut off, leaving a 12 mm ($\frac{1}{2}$ in.) flap and a 50–75 mm (2–3 in.) long exposed panel. With latex exuders the budwood should have been collected at least 4 hours previously. Incisions 50–75 mm (2–3 in.) long are made on either side of the bud about 8 mm ($\frac{1}{3}$ in.) apart and cut above by rolling the stick against the knife. A similar cut is made 50–75 mm (2–3 in.) below this, the patch gripped between knife and thumb and gently pulled off without bending it. Bruised or otherwise damaged patches should be discarded. The healthy patch, held as indicated (Fig. 43) is inserted, gripped by the flap

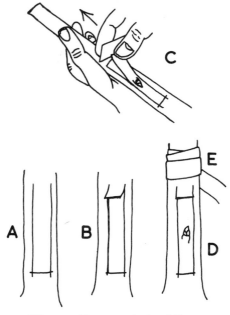

Fig. 43 Green strip budding

A. Two vertical 75 mm (3 in.) incisions and one rolling horizontal incision in the stock. B. The strip pulled back and cut off, leaving a 12 mm ($\frac{1}{2}$ in.) flap. C. Corresponding cuts outlining the bud patch. The patch gripped between knife and thumb at its lower extremity and pulled (in direction of arrow) without bending and so placed in the stock panel. D. The bud patch held by the flap and in edge-to-edge contact at its lower end. E. The plastic tie, held by two or three turns well above the patch, is spiralled downwards and half-hitched below the patch (see text).

remnant and in close edge to edge contact with the lower cut on the stock. Binding should begin a little above the patch and flap and proceed spirally downwards with adequate overlap, taking care by finger pressure that the patch does not slip sideways whilst binding. Bud-grafted plants should be cut to about 75 mm (3 in.) above the bud some two weeks after budding.

'TAILOR'S GOOSE' BUDDING

The name describes the appearance of the completed graft (Fig. 44, C). The method is advocated[104] for the working of wild olives in Algeria. The bud patch (A) includes a pair of buds. Two horizontal cuts are made in the stock (B) and connected by one vertical cut. The stock rind is cut to expose the buds when it is folded over the patch (C). Firm tying with tape, avoiding the actual buds, completes the operation. When a union has formed, a girdle of rind is removed from the stock immediately above the patch to force the bud.

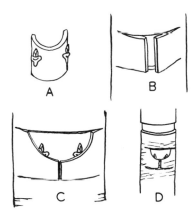

Fig. 44 'Tailor's goose' budding

A. Bud patch or ring containing two eyes which were opposite when on the shoot. B. The stock rind cut and lifted. C. The patch inserted and the stock rind trimmed to expose the two eyes. D. A union has formed and a girdle of rind is removed from the stock to force the bud.

'WINDOW' BUDDING

A somewhat cumbersome modification of both patch and shield budding. The stock must be large and the rind pliable.

157

When the rind lifts readily, a small patch (Fig. 53, B) is removed from the stock to accommodate the scion bud and permit its development. A much larger, basally hinged patch, containing the 'window', is raised and the bud shield, chip or patch is placed so that only the actual bud is visible. Firm taping, close above and below the bud, completes the operation.

CHIP BUDDING (*plate budding*)
There are many variations but it would seem that none of the more complex techniques is superior to the very simple chip method which perfectly provides good cambial contact and is easily anchored by simple tying or taping. Since the simple method now to be described is a cleft graft, it may be used over a longer season than a rind graft[81] and, provided the scion chip is properly sealed against drying, it may succeed at almost any time. It is an ancient method but only recently[68] has it been proved superior in performance to commoner techniques including shield budding with or without T incisions, and is now being used for a wide range of species of fruit, ornamental and forest trees. Those who are not yet acquainted with the method would do well to try it.

The stock is wiped clean, as with other methods. Standing over the stock, with the bud-wood in hand, a cut is made downwards in the side of the stock with the knife held horizontally so that the bottom of the cut, though curved where it begins, is horizontal at its base[67] (Fig. 45, A). A second cut (B), 25–50 mm (1–2 in.) long, judged to fit the bud shield, is made downwards to meet the first and the piece of stock so released is discarded. The bud-wood is held with its base towards the operator and is given a horizontal cut about 12 mm ($\frac{1}{2}$ in.) below the bud, and a second cut starting about 25 mm (1 in.) above the bud to join the first, thus releasing a bud chip (C) to fit the prepared stock (D). The chip is temporarily held by the stock flap (E). The stock rind is usually found to be somewhat thicker than the chip rind and a margin or layer of the outer rind of the stock should be left visible to ensure that the cambia or inner edge of the rind of both stock and chip coincide (F). Some operators prefer to cut the scion-bud first and then to cut the stock to fit the scion.

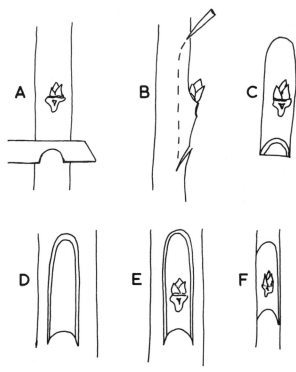

Fig. 45 Chip budding

Very firm tying is particularly important for this cleft method, along with perfect sealing with plastic or waxed tape. Twine or thread ties should be lightly sealed with a coating of wax. It may prove inadvisable to tie over the actual bud, particularly when it is very prominent; otherwise complete covering with thin plastic strip such as 25 mm (1 in.) wide 200 gauge (0·008 in.) polyethylene may improve bud take.

The chip budding of grape vines in California[113] is a development of the normal technique. The season recommended is from early August to the end of September. The bud is tied with 3 mm ($\frac{1}{8}$ in.) rubber grafting-tape. The budded portion of the stock is then covered with 150–200 mm (6–8 in.) of moist soil to protect the bud until the following spring. The stock vine is pruned of excess growth when dormant. In spring the mound of soil is removed and the stock cut through an inch or so above the bud and any scion-roots formed are cut off. The rubber tie is cut below the bud, but not removed, and the soil replaced so that the bud is just covered. In heavy soils a tube of thin waxed card placed over the bud and filled with friable compost or sand encourages straight growth. The new shoots from the 'eye' should be reduced to one when less than 300 mm (1 ft.) high.

Chip budding in the cotyledon stage. Because the simple chip-budding technique described above is readily accomplished by smooth, flat cuts with a razor or thin, sharp knife, it is excellent for grafting soft, active tissues. Provided the tissue has developed beyond the 'watery' stage, so that it maintains its structure when cut and placed together, the earlier it is grafted the sooner union will take place, given reasonably adequate horticultural attention, notably shading and irrigation. Indeed, whenever it is found that tissues of woody plants are slow to unite, or when united the bud is slow to develop an adequate shoot in a reasonable time, it is worth trying bud grafting at a much earlier stage in the life of both the stock and the scion-shoot which is to provide the bud. Examples of this are the 'pink stage' budding of mango[84] and the green chip budding of guava[71]. Costs of production of bud-grafted mango trees were markedly lowered by grafting three-week-old seedlings still in the succulent red stage with buds of selected cultivars taken from terminal shoots in the last stages of pink stem, the shoots being no thicker than 9 mm ($\frac{3}{8}$ in.). Thicker material gives too large a chip for use on young stocks.

Some two weeks prior to grafting the leaves are closely cut from the bud-wood shoots, leaving only two leaves at their tips. This remaining piece of petiole will absciss and the bud will swell. Each shoot should yield from five to eight usable buds. The normal chip-budding technique (Fig. 45) is used, taking particular care not to bruise the tissues. Workers in Florida[84] recommend wrapping the graft completely with one and a half turns of 50 mm (2 in.) wide very thin polyethylene and binding over this with rubber strip. Two weeks later the binding and covering is removed and one week later still the stock is cut down to three leaves above the graft. When the bud has grown 75–100 mm (3–4 in.) the head of the stock is removed close to the bud and the wound is sealed. The season for budding at this early stage is dependent on the availability of germinating seed.

Modified chip budding. This method is used in spring before the stock is in active growth. The rind of the stock is sliced off with the minimum amount of wood. This slice may be 25–50 mm (1–2 in.) long, the width being decided by the diameter of the stock. A similar slice, but including a bud and, of necessity, a certain amount of wood, is cut from the scion, applied to the stock, tied and sealed.

Inlay chip budding. (*See* veneering with green shoots.)

PRONG BUDDING
This method is used when it is difficult to find single vegetative buds, as when collecting scions from ancient trees. By using a short spur or prong, bearing numerous buds, there

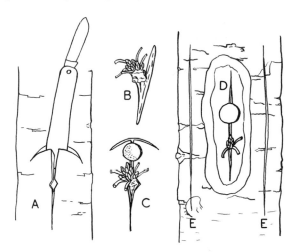

Fig. 46 Prong budding

A. The incised rind of the stock is lifted by the spatula end of the knife. B. The spur or prong cut from the scion wood. C. The prong inserted and the stock rind held down by a large-headed nail. All incisions should be sealed. D. Thick rind pared down before inserting the prong in a long vertical slit. E. Slits in the stock rind to release tension in the region of the graft.

is the best chance of obtaining new shoot growth. Prong budding is also useful in furnishing bare parts of the tree with fruiting laterals or spurs. The method has been used[70] with success in grafting ornamental flowering cherries upon *Prunus avium*. Where short bud sticks are received in a dry condition prong budding affords one of the most practical methods of saving the variety. The rind of old stocks may be pared down to half its thickness to make it more pliable. It is also an advantage to cut away small portions of the rind so that it lies snugly around the base of the prong (Fig. 46). The completed graft should be tied and sealed. Where the stocks are very large it is preferable to fix the rind with large-headed nails.

AFTERCARE OF BUD GRAFTS

The stock should be maintained in a healthy state by good cultivation, particularly in nurseries, for growth from the bud largely depends upon the vigour and health of the stock. The growth of the stock above the bud should not be reduced by pruning until the time comes to force growth from the scion.

Most tree species exhibit a powerful apical dominance so that growth from the newly-placed scion-bud is suppressed. Only by removing or checking the upper dominating growth can the scion be forced into activity and so become the trunk and branches of the new tree. Buds inserted in the early part of the growing season may be brought into growth that same season by removing, usually gradually, the leafy shoots of the stock. With small stocks, easily bent, it is often sufficient to bend the stock so that the bud is at the top of the curved stem. On a small scale, rods or iron pipes may be laid on the bushy heads of the stocks to hold them in position. When the scions have made good growth, and have well-expanded leaves, the stocks are cut through immediately above the buds. This wound may, with advantage, be smeared with sealing material to prevent drying of tissues close to the scion.

New buddings may be forced by removing a ring of rind from around the stock immediately above the inserted bud. Half-rings on the budded side often attain the same end without destroying the upper part of the stock.

The luxuriant shoots on fan-trained peaches and nectarines, which are often barren, may be furnished with fruiting laterals by budding and temporarily constricting the stock in the following manner[76]. As early as practicable in June, buds are inserted and the stock above is bound with raffia or soft twine. The bud is then tied in the normal manner. When the bud has taken it is unbound, but the upper tie is left until the bud has grown out about 100 mm (4 in.), when it too is removed.

Normally, buds inserted during one growing season are not encouraged to grow out until the following season.

Tying materials which do not break or expand as the stock thickens should be cut or removed before severe constriction occurs. With some subjects it is necessary to retie to prevent the stock rind from curling back from the scion. This retying need not be so thorough as in the first instance, so long as it retains the rind in position.

Stocks containing buds which have remained dormant until the winter are usually cut down to the bud soon after midwinter. If the stock is cut some 75–100 mm (3–4 in.) above the bud, this snag will serve as a stake to which the new growth is tied to keep it upright. This practice is described in Chapter VII. Such snags are cut off close to the scion at the end of the first growing season. The cutting off of the snag is commonly known as 'snagging'.

Though it is sometimes necessary to support the growth from a graft until a strong union is formed (see page 229), excessive staking should be avoided. Usually growth from bud grafts can safely stand alone, growing stronger thereby, and some other types of graft can well be left unsupported. It has been demonstrated that non-supported plants strengthen themselves against the stresses they meet[13,45]. The naturally-grafted pine trees, shown in Plate 12, have an interesting trunk section. The trees are firmly braced together and cannot move in line with each other below the brace but only sideways. Sections of the trunks are shown diagrammatically (Fig. 47). The sections below the junction are elliptical but those above are circular. In fact, the trunk of the larger partner is seen to be slightly greater in diameter above the junction than the minor axis below. A similar effect may be seen in a beech hedge. The main stems of the individual

12. Natural grafting

*Two pine trees (*Pinus sylvestris*) linked by natural grafting. Note the increased size of the small tree above the junction.*

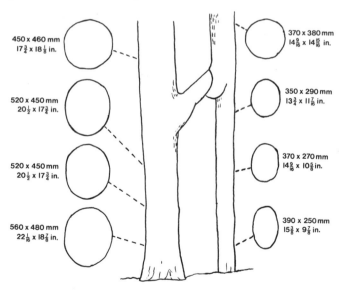

450 x 460 mm
$17\frac{3}{4}$ x $18\frac{1}{8}$ in.

370 x 380 mm
$14\frac{9}{16}$ x $14\frac{15}{16}$ in.

520 x 450 mm
$20\frac{1}{2}$ x $17\frac{3}{4}$ in.

350 x 290 mm
$13\frac{3}{4}$ x $11\frac{7}{8}$ in.

520 x 450 mm
$20\frac{1}{2}$ x $17\frac{3}{4}$ in.

370 x 270 mm
$14\frac{9}{16}$ x $10\frac{5}{8}$ in.

560 x 480 mm
$22\frac{1}{16}$ x $18\frac{7}{8}$ in.

390 x 250 mm
$15\frac{3}{8}$ x $9\frac{7}{8}$ in.

Fig. 47 The effect of stress upon trunk section

These naturally grafted trees (see Plate 12) are elliptical below the graft where they brace each other in one direction and circular above where they are independent. Transverse sections, equal scale, show trunk formation at the positions indicated.

'trees' are oval, with the shorter axis in line with the hedge. This is unrelated to the origin of the side branches, but arises from the greater strength needed to withstand wind pressures at right angles to the hedge.

In the spring following the winter in which the stocks are cut down, numerous dormant buds start into growth from the stock and, if left, compete with the scion. These stock growths should be sliced off or rubbed out as they appear, though some operators find that a few small stock growths are beneficial early in the spring, provided they are not allowed to crowd the growth from the bud. These sprouts, especially on very large stocks, which tend to die back if left without sufficient leaves, are definitely helpful in maintaining growth and encouraging healing of the cut surfaces. Moreover, when a few leaves develop on the snag it remains 'green', easy to cut, and the new wound readily heals.

b. Inlay and Veneer Grafting

Inlaying, as generally understood by grafters, implies the cutting out of a portion of the stock which is then replaced, more or less exactly, with the scion. It is an improvement on the cleft graft (page 252), since extensive damage by splitting is avoided and the wound is limited to the place of insertion. Inlay differs from veneer in degree only, for the term veneer should be reserved for the grafting of thin layers, usually the rind only, in place of the original, whereas inlaying involves the cutting out of both wood and rind. Inlaying does not always depend upon easy separation of rind from wood, but with veneering this separation is essential.

INLAY GRAFT

The stock must be larger than the scion. The basal end of the scion (Fig. 48, A) is formed into a long sideways wedge by two equal cuts. The stock (B) is cut through transversely and grooved by two cuts the same size and angle as those on the scion. Should the stock rind be thicker than that of the scion, the groove must be correspondingly deeper so that the cambium of the scion comes into contact with the cambium of the stock. The scion is thrust downwards into the groove, and, provided the work is accurately done, a firm anchorage is

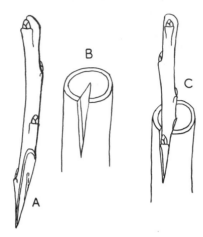

Fig. 48 Inlay graft

A. The scion formed into a long sideways wedge. B. The stock grooved by two deep cuts. C. The scion firmly inserted into the stock.

achieved. Tying or nailing is usually found to be necessary and all cut surfaces are sealed.

Inlaying is a useful method for the grafting of herbaceous and succulent plants, because exact fitting renders very firm tying or wedging unnecessary.

The careful cutting and fitting needed for inlay grafts have prevented the method from becoming popular. Special tools have been used for accurate inlaying, particularly by the French (*see* Fig. 18). One part of the tool is a V-shaped blade for preparing the scion and the other a gouge, of the same angle, for grooving the stock. Such tools make for accuracy but even so, for general work at least, the method is too cumbersome.

KERF* GRAFT

This ancient method is used to the best advantage when grafting large, woody stocks such as fruit trees with limbs 75–100 mm (3–4 in.) thick at the grafting position. The end of the limb is sawn transversely and trimmed. One saw cut (Fig. 49, A) is made for each scion to be inserted. These cuts are

* 'Kerf': the incision made by cutting, especially by a saw.

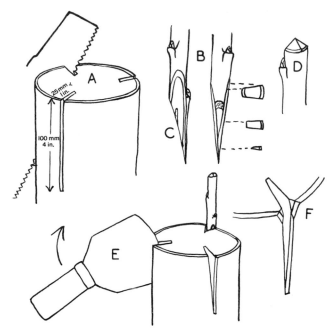

Fig. 49 Kerf graft

made very steeply so that, when finished, they extend 100 mm (4 in.) down the limb and 25 mm (1 in.) towards the middle.

Stout scions are necessary and these are prepared by two long flat cuts on opposite sides of the basal part of the scion (B), forming an angle of 18° to 20°. The inner edge of the base, often termed the heel, is sliced off (C) to permit the entry of the scion into the lower part of the cut in the stock. The apical end of the scion (D) is trimmed to a pyramid to take the blows on the harder wood in driving home the scion.

The cut in the stock is widened by removing thin slices, first on one side and then on the other. This can be done with a stout, sharp knife, but the use of a special tool (Fig. 49, E), long ago devised for this work, specimens of which have been placed by the late Dr. H. E. Durham in the museum at Hereford, renders the work more accurate. A good blacksmith or engineer can make suitable tools or they may be adapted from a cook's mincing knife[26] (see Fig. 19). The tool should be placed in the bottom of the saw cut and levered, not sliced, upwards to achieve the correct angle.

When the groove is approaching the right size the prepared scion should be tried against it and final adjustment made by cutting the scion or widening the groove. If the corners (F) of the groove are sloped back it will be easier to see the cambium whilst driving the scion. The scion is now grasped firmly in the fist and driven down by striking the top of the scion with the flat of the tool. The scion is firmly held and neither tying nor nailing is necessary. Sealing completes the operation. The apical end of the scion tends to become burred and may be trimmed afresh before sealing.

SPLIT BUDDING

This has been used for vines[16] and has been termed the 'little ship bud' because of the appearance (Fig. 50) of the scion.

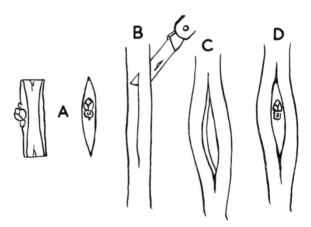

Fig. 50 Split budding or 'little ship' graft

This consists of the whole thickness of the shoot behind the bud, trimmed above, below and on either side, so that the completed scion is like the outline of a ship (A). A split is made in a shoot of the stock (B) and this is caused to gape by pushing the shoot from above (C). The scion is inserted in the split and the stock is released to grip the scion (D). Firm tying is necessary, and sealing is an advantage.

VENEER GRAFT

The scion must be much smaller than the stock and the work can only conveniently be done when the rind of the stock is readily separated from the wood. The scion-wood is cut transversely about 25 mm (1 in.) beneath the selected bud (Fig. 51, A) and shaved away down to the pith on the opposite side (B) to this bud. The scion (C) is separated from the scion-wood by a transverse cut the same distance above the bud.

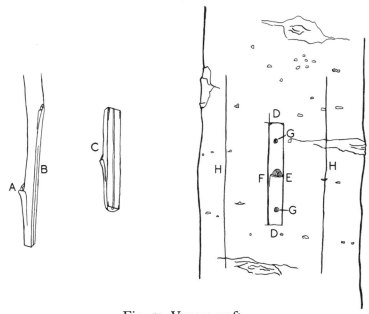

Fig. 51 Veneer graft

The scion may be lengthened to include many buds, and the apical end of these long scions may be left intact and allowed to protrude from the surface of the stock limb.

The scion is laid on the surface of the stock limb and cuts (D) are made through the rind down to the wood in line with the transverse cuts on the scion. The scion is removed and the two cuts in the stock are connected by two parallel cuts (E and F) spaced the width of the scion. The rind within the four cuts is removed and the scion is slid into place so that the flat surface is in close contact with the wood of the stock and the scion is gripped by the sides of the stock rind. The scion is held by two small nails or gimp pins (G) which are driven through the scion into the wood of the stock. Two cuts (H), an inch or two from the scion and extending a similar distance beyond it, are made in the stock rind to release the tension of the rind and to counteract shrinkage away from the scion. These two cuts are left unsealed.

Areas bare of branches may be furnished by means of this method which, in its simplest form, contains one bud only. The length of the scion may vary within the limits of the shoots available and may contain numerous buds. Long scions may be placed spirally or in line with the stock limb and are merely an extension of this simple veneer graft.

In using long scions the apical end of the scion may be left intact and allowed to protrude from the surface of the stock limb.

VENEERING WITH GREEN SHOOTS (*inlay chip budding*)
Veneering with green shoots has proved successful in propagating tea in the Dutch colonies[29]. The scion, with a bud

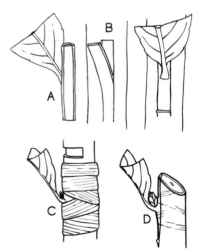

Fig. 52 Veneering with green shoots

A. Scion sliced from scion-wood, squared at each end and half the leaf removed.
B. Stock rind lifted and peeled downwards. C. The scion inserted, surplus stock
rind cut away and graft bandaged with waxed cloth. After ten days a strip of
rind is removed to force the bud D. Ten days or so later still the bud is swelling
and bandage and head of stock are removed.

midway along it, is sliced from the scion-wood and squared at each end. Half the leaf is cut away (Fig. 52, A). Two parallel cuts, just as wide as the scion and connected at the top, enable the rind to be lifted and peeled downwards (B). The scion is placed in this cavity and the peeled rind cut off level with the base of the scion-bud. The graft is now bandaged with cloth soaked in melted beeswax and any openings are smeared over. After ten days a healthy half-leaf denotes success. A strip of rind (C) is now taken off the stock above the graft. At the end of another ten days, when the bud begins to swell, the head of the stock and the bandage are removed.

PLUG GRAFT

The investigation of fusion processes is aided by the use of this graft. Potato tubers are readily plug grafted. The tubers, starting into growth[5], should be washed and then disinfected with 0·3 per cent formalin. Two square brass borers, one 15 mm ($\frac{19}{32}$ in.) square, for cutting the scion, and the other 13 mm ($\frac{1}{2}$ in.) square for the stock, are needed. A very thin and narrow right-angled knife is useful for cutting the base of the cube in each case. An eye is removed from the stock with the smaller borer and a scion eye, cut with the larger borer, is inserted into its place. It appears of some importance to place the scion in the same direction as the stock and the borer should be marked on one side and this side used parallel with the brow of the eye. The plug may be fixed by tying or pinning. All stock eyes are excised and the wounds cauterized with 8 per cent sulphuric acid. The grafts are placed in a moist medium with the engrafted eye exposed to the air, which should be kept humid.

Plants with thick rind are conveniently plug grafted by means of ordinary cork borers when the rind separates readily from the wood. It is vital to keep the scion the right way up and marking the scion or the borer is essential. One borer is used for the stock and another, one size larger, for the scion. The scion should be pressed home firmly and fixed by one or two small nails. A dab of sealing material completes the work.

VENEERING WITH RIND ONLY

When the rind is readily separated from the wood, limited

areas of the rind of mutually compatible plants may be exchanged. This technique has been used in attempts to elucidate the effect of intermediate stem pieces. A ring of rind about 25 mm (1 in.) wide is removed from one young apple rootstock, and a corresponding ring taken from another rootstock is set in its place. This scion-ring lays down new wood on the inside and rind on the outside so that eventually the tree can be regarded as double-worked. The budless scion grafts just as well as the patch in patch budding, already described. When patch or ring budding, the presence of an eye on the scion serves to indicate polarity and there is little likelihood of inverting the scion by chance but, in using rind only, as in ring grafting, there is considerable danger of inversion and the operator must take care to place the scion the right way up. It has been observed that inversion of a ring graft results in the death of the stock above the ring. If the upper or lower edge of the scion is marked by sticking a brad or small drawing-pin close to the edge or by spotting with white paint, the risk of misplacement will be reduced to a minimum.

As already remarked in Chapter II, the rind of trees in active growth separates readily in the cambial region, so that some meristematic tissue remains with the wood and some lifts with the rind. On continued exposure to the air the meristematic property of these surfaces is lost and no further growth occurs. If, however, these surfaces are carefully covered so that they are not injured by drying, they will form a callus. If a patch of rind is cut from a stock and this 'window' is covered with waxed cloth, then a new rind will form on the surface of the wood and, in due course, this rind will become continuous with the existing rind. This may be conveniently demonstrated by patch grafting a plant with purple sap upon a plant with white sap or vice versa. In order to prove that the new tissue does not arise only from the edge of the patch and spread across it, the following technique is employed. When the rind lifts readily, a patch (Fig. 53, A) is removed from a stock having white sap. A purple-sap scion is chosen and a small patch (B) of rind is removed and discarded. A scion-patch, or frame, is now collected in the normal way with the hole (B) at its centre. This scion is placed in position on the

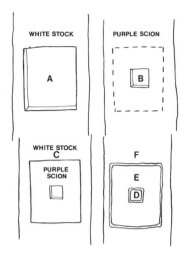

Fig. 53 Veneering with rind only

To demonstrate that new rind arises upon the surface of stripped wood, a patch of rind (A) is cut from a 'white' stock. A small patch (B) is discarded from a purple scion and the 'frame' patch is transferred to the stock (C) and covered with adhesive tape. A 'white' rind will form (D) in the middle of the purple frame (E), thus proving that healing can arise from surfaces stripped of rind.

stock (C) and is immediately and completely sealed by bandaging with waxed cloth or adhesive tape.

After a month or so the bandage is removed and the rind at the middle of the patch (D), the frame itself (E), and the stock (F) are sliced to reveal their colour. The middle, which has arisen from the stock, is white, the frame is purple and the white stock has remained white.

If the meristematic surface on wood opposite the hole (B) is killed, either by handling or drying, then the tissue at (D) will come entirely from the edges of the frame (E) and will be purple.

The fact that a new rind can arise from the surface of young wood reveals that in ring grafting it is essential to make a *complete* ring in order to avoid leaving a bridge of stock tissue which will, in cases where the scion is less vigorous than the stock, increase in size and importance. Any risk of bridging may be obviated by imposing a patch at the join in the ring after the ring has laid down a layer of wood. This is shown in the accompanying diagram (Fig. 54).

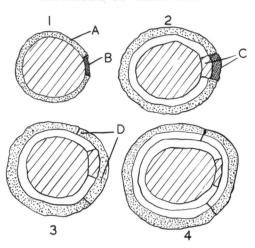

Fig. 54 Ring veneering

To ensure a complete ring the work is done in two stages. A ring of rind (A) is veneered on to the stock (B). When a layer of new wood and cambium has formed from the ring and the stock (C), a second scion (D) is placed over the gap (C). The two scions then grow in step with no risk of bridging by stock tissue.

Careful investigation[109] has revealed that the new rind on stripped wood, and also the new cambium on lifted rind, are formed in the pad of callus which arises on the meristematic surface. The new cambium develops first in the callus pad adjacent to the edges of the severed cambial ring, and this development spreads until a continuous cambium is formed across the wound (Fig. 55).

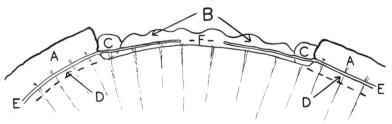

Fig. 55 Regeneration of rind on a stripped surface

Protected stripped surfaces regenerate a pad of callus (B). The cambium (E) beneath the rind (A) at the margin of the wound continues to lay down new wood (D) and a roll of callus (C) around the wound. A new cambium develops in the callus pad (B) and spreads until it joins (F) to form a continuous cambium.

c. Apical Grafting

Except for shield budding, already described, grafting apically, more particularly by the whip-and-tongue technique, is the most common method of joining plants. Under this heading come all those methods in which both scion and stock are joined at their ends as distinct from methods in which the scion is inserted in the side of the stock.

SPLICE GRAFT (*whip graft*)

This is one of the simplest methods of grafting stocks and scions of the same diameter. A long slanting cut (Fig. 56, A) is made at the basal end of the scion. A corresponding cut (B) is made at the apical end of the stock. The cut surfaces are placed together (D) so that the cambial regions (C) are in contact. When the rind of the scion is considerably thicker or thinner than that of the stock, the matching of the cambial regions brings the surfaces of the rind out of alignment, but this is of little consequence. Stock and scion are bound together and the cut surfaces are sealed.

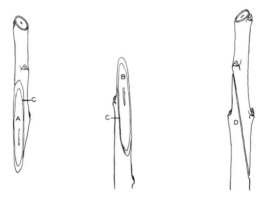

Fig. 56 Splice graft (whip graft)

A plain splice, such as this, must be held together whilst tying; hence it is not a convenient method for use at ground level, but only for bench or pot work, or high-working, where the same person does both cutting and tying. The following whip-and-tongue method does not have these limitations.

WHIP-AND-TONGUE GRAFT

This method is most suitable when the scion and stock are comparatively small, not more than 25 mm (1 in.) in diameter. It is similar to the splice graft except for the tongue which serves to hold stock and scion together, so that the operator has both hands free for tying, or so that the fitted graft can be left for an assistant to bind and seal.

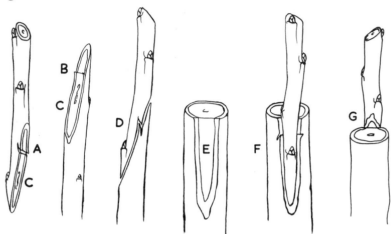

Fig. 57 Whip-and-tongue graft

Whip-and-tongue grafting (Fig. 57) is employed to best advantage when stock and scion are of equal diameter. A flat slanting cut, the length of which is about six times the thickness of the scion, is made at the basal end of the scion. A downward-pointing tongue (A) is made in the apical half of this slanting surface. A slanting cut, corresponding in length to that of the scion, is made upwards through the stock. An upward-pointing tongue (B) is made in the apical half of this slanting surface. The cut surfaces of the scion and stock are placed together so that the tongues interlock (D) and the cambial regions (C) are in contact over as great a length as possible. If the stock is considerably larger than the scion it is then advisable to take only a comparatively shallow slice off the stock (E) so that the cambium of the large stock coincides with that of the smaller scion (F) and the cambia at the top end (G) are well matched to promote a complete, non-splitting union by subsequent formation of callus. Tying and sealing of all exposed cut surfaces complete the operation. The matching of the basal tip of the scion with the lower extremity of the exposed cambium of the stock and cambial contact at the apex of the stock are important factors in the formation of a good junction. Good matching is accomplished by correct judging of the length of the cut on the scion and making only a slightly longer cut on the stock.

When beheading a thick stock to receive the scion the transverse cut is normally made horizontally (as depicted) so that the tie of twine or similar material may extend to the top of the stock without slipping, so holding the scion firmly against the edge of the transverse cut. But if wide tapes of polyethylene or waxed cloth are used, the top cut should be sloped about 45° to facilitate neat wrapping and sealing of the top of the stock.

It has been suggested that the position of the top bud on the scion in relation to the matched cambia at the graft is of major importance, but it now appears[6,15] that position of buds on scions, at least in whip-and-tongue grafting, has little or no significant effect upon either stand or growth.

Fig. 58 Whip-and-tongue graft with two scions

This method may be used to assist the healing of large stocks where operators are restricted to the use of the whip-and-tongue graft.

Whip-and-tongue graft with two scions. In grafting large stocks by the whip-and-tongue method, matching of the cambium is possible only on one side, and the healing of the large wound may be delayed. Healing may be assisted by the use of two scions (Fig. 58), but it is probably better to use some other graft, such as the strap graft, in such cases.

DOUBLE TONGUE GRAFT (*big tongue graft*)

With subjects which have a large pith this method often succeeds where others fail. The walnut (*Juglans regia*) is commonly propagated both under glass and in the open by double-tonguing selected varieties on to seedlings.

Well-ripened, solid scion-wood (Plate 13), of the past season's growth, is collected whilst completely dormant. The bases of these shoots make the best scions and, when scion-wood is scarce, it should be collected with a short length of

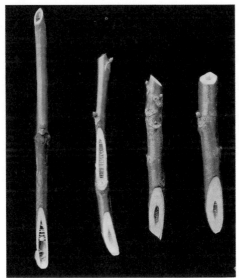

Copyright East Malling Res. Stat.

13. Choice of scion-wood

Good, solid, well-ripened scion-wood aids success and is almost essential for some subjects, e.g. walnuts, seen above. On the left, very poor 'hollow wood'; moderate material in the centre; short-jointed, good, 'solid' scion on the right.

two-year wood attached, so that this region may become the base of the scion. The wood is labelled, packed horizontally in moist sphagnum moss in boxes, which are either buried underground in a well-drained situation or placed in cold storage at 0–3°C (32–38°F). If the wood is cut into lengths for storage the cut ends should be sealed.

For grafting under glass the stocks, usually one or two-year seedlings of *J. regia*, with a diameter of about 20 mm ($\frac{3}{4}$ in.) at the collar, are lifted from the open ground in November and potted into 115 mm ($4\frac{1}{2}$ in.) 'long-tom' pots or other deep containers. Long tap-roots are shortened as necessary. The stock should stand 50 mm (2 in.) higher in the pot than it stood previously above the ground. The potted stocks are kept in cold frames until February, when they are moved into the propagation house and grafted two or three weeks later.

The scion (Fig. 59, A) is prepared with a strap (B) made by cutting upwards, from below a bud, slightly into the wood, but not to the pith. A long slanting cut right through the scion-wood is made (C) from a little above the level of the upper end of the strap to just below the lower end of the strap.

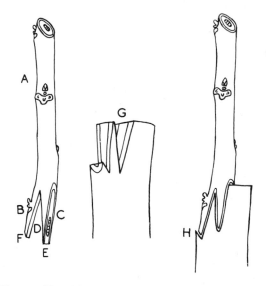

Fig. 59 Double-tongue graft (big tongue graft)

The knife is now inserted backwards in the cleft (D) and brought downwards to straighten the 'big' tongue (E). The slender tip of the strap, containing no cambium, is cut away by a slanting downward cut.

The rootstock is cut transversely 50 mm (2 in.) above the soil, and a slice is taken off the side, so that the cambium is exposed to fit the inner surface of the scion strap. A deep cleft (G) is now made, beginning 3 mm ($\frac{1}{8}$ in.) from the top of the slice and ending level with its base. The scion is now fitted and a small piece of stock rind is raised to cover the exposed cambium at the base of the scion strap (H). The graft is tied firmly with stout twine and all cut surfaces are carefully sealed.

The completed grafts are placed in closed cases with bottom heat of 21–24°C (70–75°F) until a union is formed and growth begins. Air is then admitted gradually and the successful grafts are hardened off for planting in open ground during the summer.

SPLIT-SCION GRAFT

The splitting of scions to make two scions instead of one is occasionally used in veneer grafting on large stocks, but this method is more common in propagation of certain herbaceous or semi-herbaceous plants.

Named varieties of clematis are commonly propagated by grafting upon whole or piece roots of *C. vitalba* or *C. viticella*. To avoid any risk of the rootstock suckering it must be grafted at the hypocotyl or below. The plants providing the scions are established in pots and forwarded in a warm house from early January onwards. The shoots must be firm, neither hard nor succulent. Two scions are made from each node by splitting the stem. Each scion (Fig. 60, A) is about 25 mm (1 in.) long with a leaf attached. This leaf may be reduced in size by a cut (B) above the first pair of leaflets. A cut is made at the base of the scion (C). The rootstock is cut square across and prepared by a downward cut (D), the same length as the scion, and a second cut (E) joins the first (*but see* Tailed-scion grafting, below). The scion is placed in position and tied with very thin, wetted raffia (F). No sealing is necessary. The grafted plant is immediately potted in a small pot, with the scion-bud

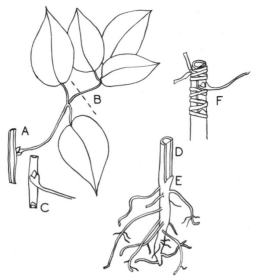

Fig. 60 Split-scion grafting the clematis

just below soil level, and placed in a closed case until a union is formed and growth begins, when the new plant is gradually hardened off and moved into a larger pot as required.

This technique has proved successful with a large variety of plants.

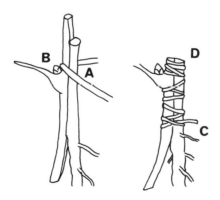

Fig. 61 Tailed-scion grafting the clematis

First turn (A) of the tie immediately above the scion-bud (B) and proceeding downwards, finishing with a half-hitch (C). Top of stock and scion trimmed close (D) above the tie.

181

TAILED-SCION GRAFT

If the split scion (see above) is lengthened downwards so that it extends an inch or so below the cut surface of the rootstock when grafted (Fig. 61), it forms a tailed-scion graft. Scion rooting is thus encouraged. When grafting clematis and some other subjects this modification of the split graft is often preferred.

MODIFIED RIND GRAFT (*du Breuil's method*)

This is an excellent method where the stock is somewhat larger than the scion, and it has been suggested as suitable for grafting the walnut and the olive. An upward cut (Fig. 62, A) is made on the scion somewhat below and on the opposite side to a bud, penetrating about one-third through the scion. A portion of the scion-wood is cut away up to this cut, in order to allow the knife to be turned, and the major scion cut (B) is made downwards through the scion. The extreme base of the scion (C) is removed to facilitate the entry of the prepared scion beneath the bark of the stock. A very thin line of rind (D) is cut away from the side of the major cut (B), just exposing the cambium and producing a straight edge to fit against the unraised rind of the stock. The stock (E) is cut

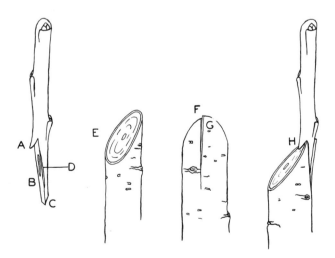

Fig. 62 Modified rind graft (du Breuil's method)

182

obliquely. The rind receives a vertical cut (F) somewhat to the side of the top edge of the oblique cut. The stock rind (G) is raised only on one side of this cut. The scion is inserted beneath this raised rind so that the lip (H) raised by the first cut (A) on the scion fits the top of the stock and the trimmed edge of the scion comes in close contact with the edge of the unraised stock rind.

Bandaging with adhesive tape and sealing of exposed cut surfaces complete the operation. Owing to the slanting cut surface of the stock, tying with non-adhesive materials is difficult.

SADDLE GRAFT

The stock and scion should be about the same size. The scion is prepared by cutting upwards through the rind and into the wood on opposite sides of the scion piece. The knife should penetrate more deeply into the wood as the cuts are lengthened and, before the knife is withdrawn, turned as sharply as possible towards the middle of the scion, so that the middle portion may be removed and the scion (Fig. 63, A) have the required saddle formation. The stock (B) is cut transversely and receives two upward cuts (D) on either side, sufficiently

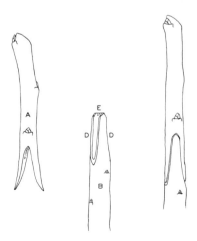

Fig. 63 Saddle graft
183

deep to expose the cambia in such a position that they may match those in the saddle of the scion. The apex (E) of the stock is shaped to fit the saddle.

Stock and scion are fitted together and firmly tied. Except where these completed grafts are placed in closed cases, all cut surfaces should be sealed.

Named varieties of rhododendrons are commonly propagated by saddle grafting using *R. ponticum* for the rootstock (*see* page 242).

Modified saddle graft. When the stock rind is readily lifted the two 'legs' of the scion may be inserted under the rind of the stock (Fig. 64). To accomplish this modification, the 'legs' must be longer than the pieces sliced from the stock. This method increases the amount of cambium in contact but has little advantage over the ordinary saddle.

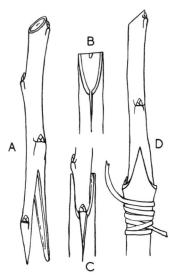

Fig. 64 Modified saddle graft

A. The scion with deep saddle base. B. The stock apex showing one side incised at the base of the upward-pointing wedge. C. The 'saddle' scion placed over the wedge with the 'legs' inserted under the stock rind. D. Side view showing both legs under the rind and binding half done.

Strap or bastard saddle graft. The scion is prepared by first raising a strip (strap) of bark (Fig. 65, A) with a small amount of wood upon the scion-wood opposite a bud (B). The knife is reversed and a slanting cut as long as or longer than the strap is made through the scion-wood to a point 25 mm (1 in.) or so below the bud. A tongue (C) is raised midway along this second cut surface.

The head of the stock is removed by a long slanting cut (D), and a second but shallower cut (E) is made on the opposite side to form an upward-pointing, unequally-sided wedge. A tongue is raised upon the surface of the first cut on the stock to correspond with the scion tongue. Stock and scion are fitted together (F), firmly tied, and the cut surfaces are sealed.

Where the stock is much greater in diameter than the scion, as in top-working established trees, described in Chapter VIII, the strap is made correspondingly longer and the tongue is usually omitted, as in the Somerset saddle graft.

Somerset saddle graft. If the tongue is omitted from the above bastard saddle graft (Fig. 65), the graft is known, at least in England, as the Somerset saddle.

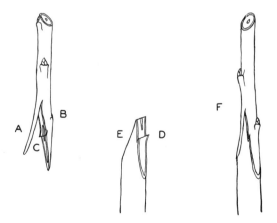

Fig. 65 Strap or bastard saddle graft

WEDGE OR CLEFT GRAFT

This is often used when scions are grafted directly on to

portions of roots. The scion may be of the same or less diameter than the stock.

The scion (Fig. 66, A) is prepared with its basal end in the form of a wedge. The stock is split at its apical end (B) and the scion inserted so that the apical portions (D) of the cut surfaces of the scion are just visible and at least one side of the cambia of stock and scion is in contact.

The graft requires tying and sealing, except where it is possible to obtain a union by placing it in moist material.

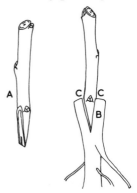

Fig. 66 Wedge or cleft graft

Wedge grafting is a convenient method for grafting herbaceous stems. The potato (*Solanum tuberosum*) (Fig. 67 and Plate 14) is wedge grafted when shoots are 150–225 mm (6–9 in.) high; this allows for tuber production and avoids risk of scion rooting. With tomato scions on potato stocks it is possible to have crops of both potato and tomato in the same season (Plate 14). When working with *Solanaceous* plants it should be noted that many solanums contain very poisonous alkaloids which are translocated across graft unions[77,83]. Tomato plants grown on thorn apple (*Datura stramonium*), tobacco (*Nicotiana tabacum*), deadly nightshade (*Atropine belladonna*) or henbane (*Hyoscyamus niger*), contain poisonous substances which may be ingested by those who eat the tomato fruits. Though eating small quantities may not prove fatal, the possibility of serious consequences must not be ruled out. There is no evidence of ill effects from eating tomato fruits produced on tomato/potato plants.

Fig. 67 Wedge grafting the potato

The cuts are made with a razor or thin, sharp knife to avoid bruising the delicate stems.

14. Tomato wedge-grafted on potato

Lycopersicum esculentum *on* Solanum tuberosum *yielding two useful crops.*

The tubers to be used as rootstocks should be sprouted in full light to obtain firm tissue for grafting. An exhaustive survey of potatoes in 1926 revealed that the stem of 'Eclipse', unlike all other varieties, is solid in cross-section throughout its whole length, there being no medullary cavity. There may now be other varieties of similar nature. Young tomato seedlings form excellent scions but discarded axillary shoots from plants grown early in the season under glass are quite adequate. Scion and stock shoots should be similar in size and condition. A safety-razor blade or a very thin knife facilitates the preparation of succulent shoots for grafting. Tying with raffia completes the work. Should the scion tend to be forced out of the stock cleft in the process of tying, a strand of the material passed over the base of the lower petiole will retain it in position. The grafted plant should be placed in a moist atmosphere to prevent wilting, which may occur until a union is effected. Sealing of the exposed cut surfaces with petroleum jelly or other substance is not necessary.

Wedge grafting in the cotyledon stage. Wedge grafting is also highly successful with young seedlings of all kinds when they are in the cotyledon stage. For example the glory pea (*Clianthus dampieri*), which grows poorly on its own roots, is grafted upon *Colutea arborescens* raised from seed. When the young seedlings have developed two or three true leaves the stem is cut transversely immediately above the cotyledons and cleft between them. An equally young, or younger, plant of the scion variety is cut off at soil level and made into a wedge below the cotyledons. This prepared scion is inserted in the cleft stock and tied in position. The grafted plant is kept in a closed case until a union has formed.

Apples are sometimes, for experimental purposes[42], grafted in the early cotyledon stage. If the seedling plant providing the scion is to be kept, the scion is collected by severing immediately above the cotyledons to permit bud development from their axils. If the scion plant is not required it is best to include the cotyledons in the scion. Unfolding scion leaves are trimmed to reduce transpiration (Fig. 68, B), the young shoot is laid on a clean surface, and a thin slice of tissue is removed from either side, using a razor blade, to form a

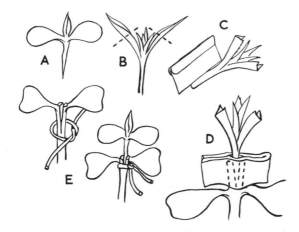

Fig. 68 Wedge grafting in the cotyledon stage

A. Scion including cotyledons. B. More advanced scion trimmed to reduce transpiration. C. Cut down on to a clean surface to form wedge. D. Graft secured with self-sealing rubber (see Fig. 69). E. Similar graft tied with thin rubber strip or thread.

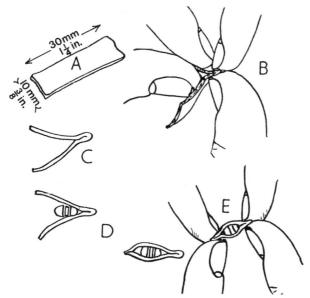

Fig. 69 Use of self-sealing crêpe rubber

A. Strip of self-sealing crêpe rubber. B. Folded and nipped along folded edge between thumb and forefinger. C. Folded strip ready for use. D. Strip placed around the graft and free ends in contact. E. Strip tightened by nipping both sides equally between thumbs and forefingers.

wedge. The rootstock is beheaded about 6 mm ($\frac{1}{4}$ in.) above the cotyledons and cleft at right angles to them, the cleft extending down to cotyledon level. The scion is inserted and sealed with self-sealing crêpe rubber (Fig. 69). The grafted plant (Plate 15) requires the protection of a closed case for about ten days.

E. Yoxall Jones   Copyright East Malling Res. Stat.

15. Grafting in the cotyledon stage
(Left to right) A rootstock beheaded and cleft; the scion fitted; tied and sealed with self-sealing rubber; a grafted plant six weeks later.

Arabica coffee (*Coffea arabica*) has been grafted on robusta coffee (*C. canephora*) in the early cotyledon stage in Guatemala to control nematode infection[105]. The use of *C. canephora* as a rootstock not only controlled nematode infection but reduced damage by *Rhizoctonia solani*. Seed of the rootstock is sown five days earlier than that of the scion. At the 'matchstick' stage, with the cotyledons still enclosed, the hypocotyl of the scion is cut to an 8 mm ($\frac{1}{3}$ in.) wedge and the rootstock beheaded and cleft below its cotyledons to receive the scion. Thin polyethylene strip has been used for tying; stericrêpe (page 135) might prove more convenient and less likely to damage the delicate tissues. Place in a closed case for a week, then harden off for transfer to the nursery via plastic or fibre containers.

The use of germinating seed to aid regeneration of plant

parts is an ancient idea[40]. Recently[72,93] large seeds have been used with good success to nurse shy-rooting cuttings of related species. Seeds which retain their cotyledons below ground (hypogeal) are germinated and, when the radical is some 50 mm (2 in.) long, both radical and plumule are sliced off to reveal the cotyledon connective. This connective, or petiole stub, is split with the point of the knife and the wedge-shaped base of the scion is inserted so that the cambium of the scion is in firm contact with the cut surface of the petiole stubs. The seed grafts are set 50–75 mm (2–3 in.) deep in an open medium, under protection, as are non-grafted leafy cuttings.

The 'stone grafting' of mango is said to have a number of advantages over other methods. The method is essentially as described above. Both wedge and splice techniques are successful and the whole operation is performed under a lath house, or other improvised shelter, with the help of a nursery frame for graft establishment. In this environment union and sprouting occurs in some two weeks, when the plants are gradually hardened off for later transference, in containers, to the open field.

Macadamia has also responded well to seed grafting. The technique is much as described above, but it is recommended that the scion-wood, preferably from non-flowering shoots, be cinctured at least four weeks before collection. The scion should have two whorls of buds and should be de-leafed at the time of collection. Grafting is done when the radicle has extended about 50 mm (2 in.), by which time the shell is sufficiently opened to accept the scion. The wedge-based scion is inserted by forcing it into the cotyledons as far as possible without splitting the nut into two pieces. The insertion must be made at an angle of 90° with the cotyledon faces, not into the division. The top of the scion should be sealed and the plant set in a well-drained medium to avoid rotting. Winter or spring grafting is recommended, the seed stocks having been forwarded as required.

Wedge grafting the growing-point. A number of plant viruses are not fully systemic, particularly when the plant is growing rapidly, for then the invasion of new tissue by the virus may lag behind, leaving the growing-points temporarily free of contamination. It is thus possible to obtain healthy progeny from a diseased parent. Preferably the parent should be grown as rapidly as possible, free of contamination by insect vectors or other carrier agents. Growing tips are, inevitably, very tender and are best grafted on to fairly tender stocks, which must be uncontaminated seedlings or plants known to be completely free of the virus concerned.

The method is very like wedge grafting in the cotyledon stage (Fig. 68) with the scion as small as possible. The healthy stock, which should be in rapid growth, is lightly tipped and cleared of leaves within 25 mm (1 in.) or so of the cut end (Fig. 70, A) and cleft by a razor or surgeon's knife. A tip scion not more than 12 mm ($\frac{1}{2}$ in.) in length is formed into a wedge (B); cutting lengthways on to a card is sometimes easier than in the hand (C). The scion is inserted (D) and held with self-sealing rubber as described for the cotyledon method (Fig. 69). The scion leaves, still folded, should not be bruised or cut, as this may lead to rotting. A sleeve of thin polyethylene (25 gauge, 0·001 in.) is put on (E) and the grafted plant returned to warmth for a week or so, when the sleeve may be opened and then removed.

The rate of shoot extension, prior to scion collection, may be increased considerably by a light spray of an aqueous

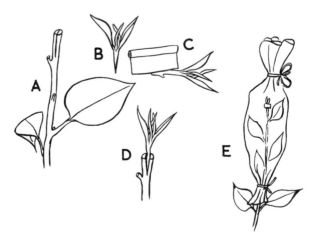

Fig. 70 Wedge grafting the growing-point

solution of gibberellic acid. Two applications, a week apart, of some 2 ml (0·07 fl. oz.) of solution, containing one ten-thousandth part of the acid, has proved effective.

The transmission of virus diseases in the hop (*Humulus lupulus*) has been successfully accomplished by means of the wedge graft. The stocks, raised from cuts, are grown in pots and grafted, whilst in the young herbaceous state, with scions of terminal growths (Fig. 71). A razor or surgical knife is used to form the base of the scion into a wedge (A). The stock is cut transversely between the second and third nodes and a wedge of stock tissue is cut out (B) about 12 mm (½ in.) deep. The fitted scion is fixed by tying with thin surgical rubber strip, and no other sealing is necessary.

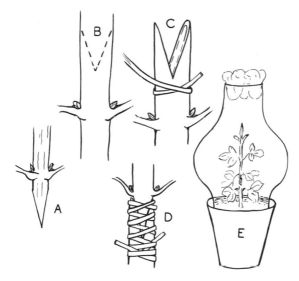

Fig. 71 Herbaceous wedge graft

All cuts are made by razors. Succulent scions may be laid on a clean board whilst cutting. A. The base of the scion formed into a wedge. B. The stock (potted) prepared by two cuts. C. Thin rubber strip given one turn on the stock before inserting the scion. D. Bind upwards and then downwards finishing well below the graft by tucking under the last turn. E. The graft temporarily protected by a lamp glass plugged with cotton wool.

The newly grafted plant is kept in a moist atmosphere until satisfactorily united.

Petiole wedge graft. Used by virologists for indexing and artificial infection of strawberry plants in the field[10], this method might well be adopted for similar studies with a wide range of plants. Mature, fully formed leaves are used (Fig. 72). The lateral leaflets are removed from the scion and some two-thirds of the remaining central leaflet is cut away to reduce transpiration. The petiole of the scion is cut to form a wedge 10 mm ($\frac{3}{8}$ in.) long (A). The central leaflet is removed from a leaf on the stock and the petiole is cleft (B) to receive the prepared scion. A wrapping of self-sealing crêpe rubber (*see* p. 135) is put on (C) and the grafted plant is lightly shaded for about a week until a union is established.

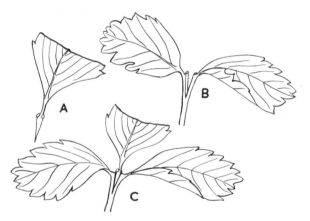

Fig. 72 Petiole wedge graft

Oblique wedge graft. This graft can take the place of the popular whip-and-tongue method in favourable circum-stances. The scion is prepared with its base in the form of an unequally-sided wedge (Fig. 73, A). The scion should be as short as practicable and it is usually necessary to prepare the base of the scion before separating it from the scion-wood in order to grip the scion whilst cutting. The apical bud (B) and longer basal cut should be on the same side of the scion. The stock is cleft in the side (C), no deeper than the middle, and is pressed aside to open the cleft (D). The scion is inserted with its longer cut towards the stock and the apical bud uppermost.

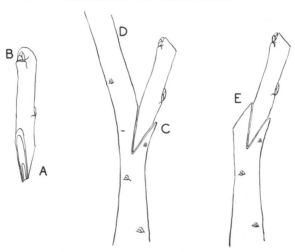

Fig. 73 Oblique wedge graft

The stock is then released and cut off close above the inserted scion (E). The graft should be well sealed with a hard-setting wax. This graft needs no tying.

Strapped wedge graft. This method provides ample cambial contact and quickly results in a mechanically strong junction. It is more easily accomplished with soft-wooded plants and is excellent for most herbaceous subjects. Walnuts and other somewhat difficult subjects for grafting have responded to this treatment.

The method with walnuts is to have small seedling rootstocks established in pots with the upper portion of the taproot exposed above the rim of the pot. Scions are collected when the wood is half-ripe, or just beginning to lignify, and the pith is beyond the very 'watery' stage. The scions may be either the terminal portion of the shoot, or pieces with one or two eyes, with leaves attached. Half of each compound leaf is cut away to reduce transpiration (Fig. 74).

The woody stem of the stock is almost entirely removed and the stock split (A) down the middle. A thin slice (B) of rind is pared from both sides of the stock.

A strip of rind is cut upwards on either side of the base of the scion (C) and the middle portion is cut to a wedge which is

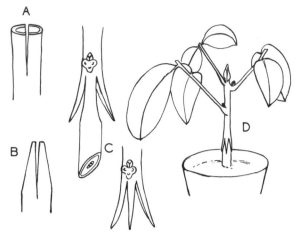

Fig. 74 Strapped wedge graft

A. The stock cut transversely and cleft. B. The stock sliced upwards on either side. C. Strap cut upwards on either side of the base of the scion and the middle formed into a wedge. D. Scion fitted to stock ready for tying.

then fitted into the cleft in the stock and the two scion straps are folded down on to the pared areas on the stock. Firm tying completes the work and the grafted plant is placed for two or three weeks in a closed frame to keep the leaves from wilting. In some cases sealing the cut surfaces is considered advisable. Melted paraffin wax is suitable.

GAP GRAFT (*Pfarrer Dees's graft*)

This is an ancient and somewhat complicated, but interesting, method described to the writer by the late Dr. H. E. Durham of Cambridge. Both scion and stock must be approximately the same size and the stock rind lift readily. The stock is cut transversely and the rind slit (Fig. 75, A) down either side for 75–100 mm (3–4 in.). The stock rind is peeled down in two halves (B) and the bared wood is cut through about 25 mm (1 in.) above the stripped rind in a V or wedge form (C), with sides as steep as practicable, at right angles to the slits (A) in the rind. A cleft 12–20 mm ($\frac{1}{2}$–$\frac{3}{4}$ in.) deep (D) is made, continuing the base of the V downwards.

The scion, which may be a whole shoot, is made into a wedge (E) at its base to fit the V, and extends in a long-drawn-

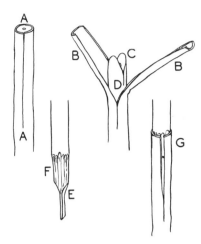

Fig. 75 Gap graft (Pfarrer Dees's graft)

out point to enter the cleft in the stock. The rind (F) at the base of the scion is scraped off for a distance sufficient to receive the lifted stock rind. This scraping must not go deeper than the cambium. The prepared scion is now firmly pushed home into the stock and the rind is folded upwards (G) to cover the junction and scraped parts. Tying proceeds from below upwards, and sealing completes the operation.

ABUT GRAFT (*'end-to-end' graft*)
This has been used for various research purposes, and a description of the technique may serve to encourage others to employ this graft where appropriate.

The method is sometimes used to join the style of a paternal parent on to the ovary of a maternal parent to overcome sexual incompatibility between two species or varieties[133]. A single wire support based on the pedicel is continued upwards close alongside the ovary and stigma. The maternal style is cut off and the paternal style is shortened, according to need, and placed on the ovary, to which it is glued with gelatine. The style is bound to the wire support by a few turns of spider's thread.

A simpler technique[12] has been used for shortening and splicing styles. The styles in almost-mature flower buds are

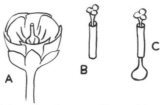

Fig. 76 Abut ('end-to-end') grafting of styles

exposed and shortened by a transverse cut across the un-opened flower (Fig. 76 A). A stigma and style of the scion variety are collected, shortened, and inserted in the end of a grass straw (B). The other end of the straw is placed on the

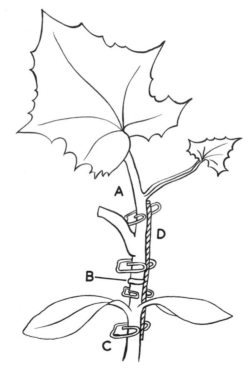

Fig. 77 Double-working cucurbita by abut grafting

A. A scion of C. sativus. *B. A thin slice of* C. melo. *C. A rootstock of* C. ficifolia. *D. A splint of wire gauze held by modified paper clips. A technique used in investigating incompatibility problems (see page 56).*

maternal style (C) and pushed down until an upward movement of the scion, relative to the straw, indicates contact of the cut ends. The abut grafting of cucumber plants (Fig. 77) was mentioned earlier on page 197.

CROWN GRAFT

The term crown in connection with grafting derives from the appearance of the completed grafts on large stocks (Plate 23). A large limb, or branch, may receive a number of scions which, being in a circle, lend a crown effect to the work[86]. Numerous methods, varying considerably in detail, are used in crown grafting. The more practical of these are the cleft, the oblique cleft, the veneer crown, and the rind methods, with their commoner modifications. All are described in Chapter VIII when discussing the grafting of established trees.

d. Side Grafting

Under this heading come all those methods where the scion is placed in the side of the stock. Most of the veneer and inlay grafts should properly be included in side grafting, but as they involve somewhat specialized technique they have been described under the appropriate heading 'Inlay grafting', page 165. Side grafting permits the stock to function more or less as it did before it was grafted until the union is complete, when the head of the stock is removed just above the junction. In some cases, notably in frameworking detailed in Chapter VIII, the stock is retained.

SIDE CLEFT GRAFT

Two slanting cuts, one (Fig. 78, A) slightly longer than the other (B), are made at the basal end of the scion, at opposite sides, to form a gradually tapering wedge. An incision (D) is made in the side of the stock at an angle of about 20° and of sufficient depth to permit the insertion of the wedge. The longer side of the wedge is placed against the stock stem in such a position that the cambial regions of stock and scion are in contact. By bending the stock so that the incision (D) is opened, the scion may be pushed home more readily. All exposed cut surfaces are sealed. Provided the incision (D) is

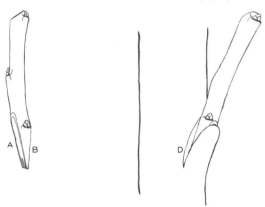

Fig. 78 Side cleft graft

made sufficiently deep into the wood of the stock, the scion is firmly held and tying (but not sealing) is unnecessary: otherwise the grafts should be tied.

Many choice shrubs and conifers are commonly propagated by side grafting to their appropriate rootstocks. Where the stocks are thin or sappy, rather longer cuts are made than described above and it is usually necessary to tie carefully with raffia or prepared strip. This work is carried out under cover and the grafted plants are usually kept under glass until well joined. There are many variations in technique according to the subject and the facilities of the nursery. Often the rootstocks are established one season in small pots and are transferred to glass-houses during the winter. Grafting takes place when root growth begins in late winter or early spring. The grafted plant is placed either in a closed case to prevent drying-out or on the open bench, as described in Chapter VII. When a union has formed the top of the stock is cut back, usually in two or more stages, and the tying material is released. By this time the plants should be ready for plunging outside in a sheltered part of the nursery.

SPLICED SIDE GRAFT (*veneer side graft*)
Small plants established in pots are conveniently grafted by this method, which is widely used. The stock stem is cleared of leaves in the region of the graft and a slice of rind, the width

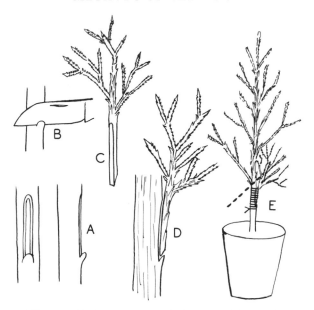

Fig. 79 Spliced side graft (veneer side graft)

A. A slice of rind containing a trace of wood removed from the side of the stock. B. A shallow cut from which the knife is withdrawn to make the straight flat slicing cut as at A. C. The scion with long cut and short basal cut on the opposite side. D. Scion fitted to the stock with base resting within small tongue made at B. E. Graft tied. When united the head of the rootstock is reduced by stages until removed entirely at the dotted line.

of the scion, with a trace of wood, is removed from the side of the stock (Fig. 79, A) by a downward cut. The lower end of this cut may terminate in a small step or tongue. The scion receives one long simple cut corresponding in length and width to that on the stock, to which it is immediately applied and firmly tied. Open-air work requires sealing but this is not necessary in humid conditions under glass. The head of the stock is removed once a union is assured (*see also* page 132).

SIDE TONGUE GRAFT

This graft is most successfully manipulated when the diameter of the scion is slightly less than that of the stock, and it can be used also when the scion is much smaller than the stock. A long, flat, slanting cut is made at the basal end of the scion. A

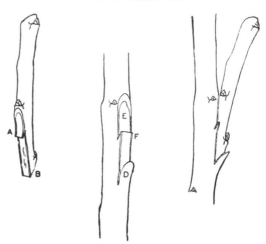

Fig. 80 Side tongue graft

downward-pointing tongue (Fig. 80, A) is made in the slanting surface. A short cut (B) is made on the opposite side to form a chisel-shaped base to the scion. An upward-pointing tongue (D) is cut in the stock to hold the base of the scion. A downward cut (E) is made as far as the tongue (D) to expose the cambial regions in such a position that they correspond to those of the prepared scion. A tongue (F) is made to interlock with that of the scion. The tongue (D) covers the chisel-shaped base of the scion (B). Stock and scion are bound together and sealed.

SLOTTED SIDE GRAFT
This method is useful when the stock is considerably thicker than the scion. The scion is prepared with its basal end in the form of an unequally-sided wedge (Fig. 81, A and B). The stock receives a downward cut penetrating to the wood. A transverse cut at the base of the downward cut produces a step (D) in the side of the stock. Two parallel cuts, very slightly closer together than the width of the scion to ensure a tight fit, are made downwards from the transverse cut. These two cuts should equal the length of the longest slanting cut of the scion. The rind (E) between the parallel cuts is raised slightly, and the scion, with its longer cut surface towards the branch,

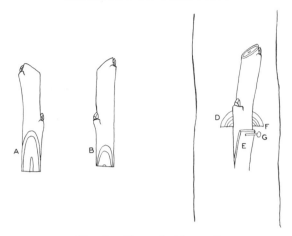

Fig. 81 Slotted side graft

is pushed down under the raised rind until the whole of the tapered base of the scion is below the level of the transverse cut (F) on the stock. A gimp pin (G) is driven through the upper part of the raised rind, through the scion and into the wood of the stock. Exposed cut surfaces are sealed.

SCOOPED SIDE GRAFT

This unusual method was devised[57] to overcome difficulties in grafting pawpaw trees (*Carica papaya*) due to a too copious sap flow, rotting of the cut surface, and a tendency of the stock to die back from the cut surface. Doubtless this method would prove efficacious with other subjects which behave similarly.

The pawpaw stock, about 50 mm (2 in.) in diameter, receives a transverse cut (Fig. 82, A) three-quarters through the stem about 150 mm (6 in.) above soil level, at which point the stem is solid; 250 mm (10 in.) above this a deep slice or scoop (B) is taken from the stem down to the first transverse cut. All cut surfaces, except the grafting surfaces placed in contact, are swabbed with potassium permanganate solution (dark red in colour when viewed in a glass tumbler). This helps to prevent rotting and checks the exudation of sap. In the step formed in the side of the stock a cleft (C) is made downwards as close as practicable to the standing part of the

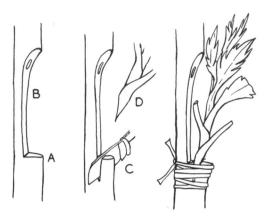

Fig. 82 Scooped side graft

stock. This cleft may be widened by removing a wedge, by means of a second cut, to reduce pressure on the soft tissues of the scion. The scions are usually side shoots 150–200 mm (6–8 in.) long with firm bases. Scions from young trees are best. The leaves are removed from the scions by severing the leaf-stalk, and the base of the scion (D) is made into a wedge by two equal cuts. The scion is inserted and tied. In pawpaw grafting the soil is then mounded up so that the apex of the scion is just buried. The stock is removed in a month or so when the scion begins to grow, the soil remaining as a protection from the sun.

SIDE RIND GRAFT

There are numerous ways of inserting scions under the rind of large stocks, more particularly where the stocks are the branches of comparatively large trees. Two of these methods, slit grafting and inverted L grafting, are described in Chapter VIII as suitable for use in frameworking. In most of the methods the scion is made into a wedge, of varying length, and the rind is lifted much or little, according to the size of the scion. A T-shaped incision in the rind of the stock is commonly used. This will receive a scion with a long taper at the base, or a heeled lateral shoot may be used as the scion somewhat as in prong budding (page 161). Another method, only of use where the stock is very much thicker than the scion, is the 'plant pot' rind graft (Fig. 83). A side rind graft for nursery use is described in Chapter VII (Fig. 97).

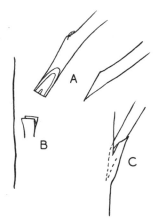

Fig. 83 'Plant pot' rind graft

A. Scion with wedged base. B. Converging incisions in thick rind of stock ('plant pot' outline). C. Scion thrust well down and held firmly beneath the rind of the stock. The graft should be sealed.

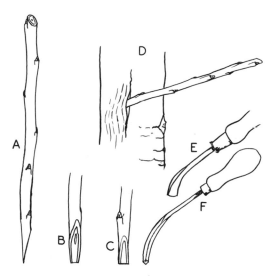

Fig. 84 Awl graft (needle graft)

AWL GRAFT (*needle graft*)

This method is used to furnish bare limbs with lateral branches[52]. The scion is prepared (Fig. 84, A, B, and C) with its base in the form of an elongated wedge of unequal sides. The larger cut (B), which is eventually placed next to the wood of the stock, may be slightly concave with advantage, as it has to fit upon the convex surface of the wood. The smaller cut (C) is made in order to enable this part of the scion to join on to the lifted rind of the stock.

A chisel-bladed tool is inserted through the rind of the stock and between it and the wood. The tool should enter from an angle of 20° to 30° above the horizontal. The tool is removed and the prepared scion inserted (D) in its place and sealed.

Various tools have been used for awl grafting, such as a small screw-driver (E) slightly curved at the tip; a sacking-needle (F) fixed in a handle and with the point ground to make a chisel end; and a large penknife blade with the point sharpened on both sides.

PEG GRAFT

This side graft is only possible where the stock is very large and the scion, leafless, is firm and strong. It has been used with moderate success in refurnishing large branches with laterals[101]. The scion is driven into the stock by gentle

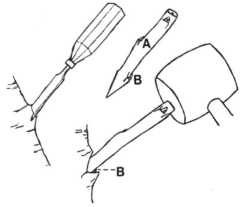

Fig. 85 Peg graft

hammering, and must be short and rigid. The scion-base is prepared in the form of a wedge (Fig. 85, A), the cuts being about 13 mm ($\frac{1}{2}$ in.) long, immediately below a bud (B). Chisels of a width to suit the majority of scions, 6 and 9 mm ($\frac{1}{4}$ and $\frac{3}{8}$ in.), are used. The bevel is doubled in length by grinding. The chisel is mallet-driven 20 mm ($\frac{3}{4}$ in.) into the stock from a little above a right angle. All the incisions are made before any scions are inserted, so that the knocking does not dislodge the scions. The prepared scions are now driven into the incisions by gentle blows with the mallet. They must be driven right home, so that the cuts on the scion are buried completely and the basal bud (B) is just visible inside the stock rind. This graft requires careful sealing and, quite often, resealing a week or so later.

LEAF GRAFT

Leaf grafting has been used with success[97] in virus transmission studies in passion fruit (*Passiflora edulis*). The leaf remains in an assimilating condition for two or three months after grafting. This technique might also prove valuable in providing extra leaves on stem cuttings which fail to root because they are ill-provided with leaves (*see* 'Nurse grafting', page 213).

A young but fully unfolded leaf is removed from the scion variety with 25 mm (1 in.) of petiole. This petiole is sliced or scraped to form an elongated wedge. In a young stock stem a downward shallow cut is made, raising a thin strip of cortex. The prepared petiole of the scion is inserted right up to the leaf-blade and bound with surgical strip rubber which holds and seals the graft. The engrafted leaves must be kept in a humid atmosphere until a union has formed, otherwise the leaf will quickly wilt and die. Leaf grafts are successful on any part of a stem regardless of the position of any existing leaves.

The petiole graft (page 194) has been used in a similar way in virus transmission studies with the strawberry.

A very simple technique has proved successful in virus transmission in a wide range of plants, both deciduous and evergreen[124]. Small pieces of leaf are dropped into sliced cuts in the rind of the stock which is then tied to prevent it opening again. A year or so after the operation, leaf pieces have been found alive and firmly united to stock tissue, but only at the edges of the leaf or at other wounds made on its surface. The leaf pieces remain alive and green even though buried by a layer of new wood three or four millimetres thick.

e. Bench Grafting

This term may be applied to any grafting process performed whilst both stock and scion are unplanted, regardless of the actual technique involved. It is termed bench grafting or bench working because it is commonly carried out on a bench.

BENCH GRAFTING ON ROOTS

In North America, where the severe winter climate prevents nursery operations at that season, apple and other rootstocks, usually seedlings, are lifted in the autumn and graded and stored in frost-proof buildings for the winter.

Grafting is done from December to March, more often by the whip-and-tongue method, though the wedge graft is also used. If one rootstock is used for each scion it is called whole root grafting, but if the main roots are separated and each receives a scion, then the method is known as piece root grafting (Fig. 86).

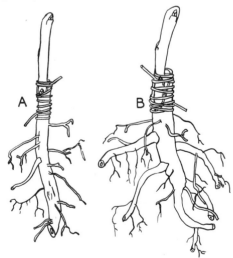

Fig. 86 Bench grafting on roots
A. Piece root grafting. B. Whole root grafting.

The rootstock stores are often termed cellars. In the grafting season enough material is drawn for each day's work in the grafting room. The parts to be grafted should be washed or wiped clean and then grafted in the chosen way. Whichever method is used, it is important to make a good join by matching the cambia, taking particular care that the cut surfaces of the scion lap on to the cambium of the stock and not on to the outer rind surface. Bad fitting often results in callus knots at the union. American and other investigators have shown that callus knots, or overgrowths, can be largely prevented by good fitting, and further reduced by the use of adhesive tape for binding. A well-fitted wedge graft has been reported as more free from this trouble than either the whip-and-tongue or the double-tongue graft, but almost any method, *provided the cambia are well fitted*, will prove equally satisfactory. In nurseries handling many thousands of bench grafts, special binding machines are employed.

After tying, sealing is rarely necessary; the grafts are bundled in fifties or less according to size, labelled with non-perishable labels, and stratified horizontally in callusing beds, or boxes, containing moist peat, sphagnum moss, or sand. The temperature should not rise above 4°C (40°F). A temperature of 2–3°C (36–38°F) is probably ideal, but an occasional drop to freezing-point is not harmful.

When the buds on the scion begin to push, or soil conditions permit, the grafts are taken from the callusing pits and lined-out in open ground, 225–300 mm (9–12 in.) between the grafts and 0·9–1·2 m (3–4 ft.) between the rows. Deep planting is the rule, only the apical bud on the scion remaining above the soil. The buried ties rot and burst before constriction occurs. Deep planting encourages strong growth and also rooting from the scion. When it is necessary to prevent scion-rooting, the grafts are planted very firmly with the lower lip of the scion just above the natural soil level. The scions are then mounded with soil so that only the apical bud is visible. When they have made good growth, usually by midsummer, the mound is worked down again by routine cultivations and the base of the scion is left just clear of the soil.

Under the best conditions, the maiden growth reaches

0·9–1·2 m (3–4 ft.) proving perfectly satisfactory. In the temperate climate of England, where rootstocks are almost always established in nursery rows and worked *in situ*, bench grafting of young apple rootstocks has not consistently given such good results.

In addition to the apple and other deciduous fruits, very many widely differing subjects are readily propagated by root grafting. Pieces of wistaria root joined to young shoots give rise to excellent plants with compact root systems. Named varieties of rhododendron are grafted upon piece roots of *R. ponticum* and the resulting plants are superior to those stem grafted, at least in their complete freedom from sucker growth from the stock. The saddle graft is used (Fig. 87). In January or early February roots are collected by digging around specimens of *R. ponticum*. The roots should be

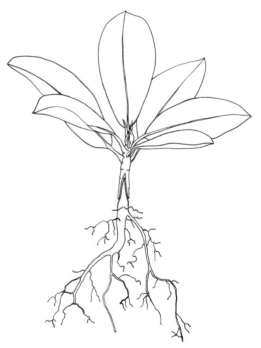

Fig. 87 Bench grafting the rhododendron
A named variety saddle grafted on to a root of R. ponticum. *This graft needs tying and the protection of a propagating case.*

approximately the same size as the scions but larger or even smaller roots can be used. The root pieces are cut into 150–200 mm (6–8 in.) lengths and kept moist. The scions, usually including the apical bud, not longer than 150 mm (6 in.) and retaining their leaves, are saddle grafted on to the apical ends of the roots and tied with raffia or tape. No sealing is necessary. The grafted plants are placed vertically close together in granulated peat in a propagating frame and in from six to ten weeks, when a union has formed, the grafted plants are potted and gradually hardened off for outdoor planting.

CUTTING GRAFTING

Plants that are difficult to propagate by either shoot or root cuttings can fairly readily be multiplied by grafting together a shoot and a root cutting which is then planted as a cutting. Certain rootstocks may be raised in this way, to be grafted later with the chosen scion varieties.

Cutting grafting is used on a very large scale in vine propagation in many different parts of the world[100]. American vines, immune to phylloxera, are used as rootstocks on which are grafted selected scion varieties susceptible to phylloxera. Mother or stock plants of the American vines are cultivated to provide suitable canes and the scions come from fruiting plants in the vineyards.

Shoots intended for rootstocks are taken from the American vines as soon as the leaves have dropped or, in some countries, up to February. These shoots are made into cuttings in lengths varying from 250 mm (10 in.) to 500 mm (20 in.) or more. The longer cuttings are necessary on the drier soils. These cuttings are stratified horizontally in moist sand in boxes or in a well-drained position outdoors. The temperature should not rise above about 7°C (45°F) during storage, which lasts until April. This keeps the cuttings in good condition without undue callusing, which otherwise uses up the food reserves of the cuttings.

The scions, collected in February or early March, are cleared of tendrils and buried 150 mm (6 in.) in soil. In April both stocks and scion-wood are washed free of grit and placed on benches for grafting. The eyes are cut from the stock and

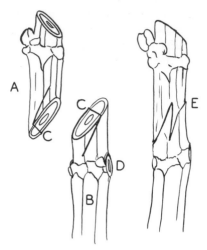

Fig. 88 Cutting grafting the vine

This graft must be accurately fitted and then requires no tying or sealing. Buried in warm, moist material callus quickly forms and locks stock and scion together.

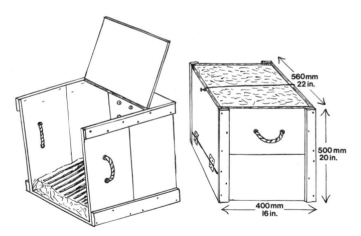

Fig. 89 Callus box for grafted cuttings

The box is placed on its side and loaded with 500 grafted cuttings. The outer packing may be sphagnum moss and the inner sawdust (conifer). When full the hinged side is closed and the box turned to bring the cuttings upright. A 75 mm (3 in.) layer of sawdust, and a covering of moss, completes the filling. The whole is heavily watered, with water at 27°C (80°F), before placing in a warm chamber.

the scion-wood is cut into pieces about 50 mm (2 in.) long, each possessing one eye. The scion is grafted on to the stock by a short whip-and-tongue method (Fig. 88). It is thought to be an advantage to place the scion so that its eye comes into the place previously occupied by the next eye above the apical eye on the stock. The scion is cut with a rather short slanting cut (A) at 45° with the direction of the scion. A similar cut (B) is made at the apical end of the stock. These cut surfaces are tongued by cuts (C) starting in the wood and crossing the pith. Scion and stock are firmly fitted with no tying or sealing. The grafted cutting is now placed vertically in a callus box (Fig. 89) containing sawdust (conifer) which is kept at a temperature of 21–24°C (70–75°F) for from two to three weeks and then at 16°C (60°F) for a further ten days. The callused grafts are now set out in the nursery ground 50–75 mm (2–3 in.) apart in rows spaced for horse culti-vation. In some districts, notably the Rhineland, special planting methods are in use. When setting out in the nursery row, the callused grafts are planted half their depth and then mounded with soil to within about 25 mm (1 in.) of the base of the scion. The scion is then completely covered with dry sawdust. This protects the scion from drying winds and discourages scion-rooting. An improved technique is also practised. In this the grafted cuttings are placed in cardboard tubes with compost around the stock, so that the cuttings are not removed from the compost immediately surrounding them but are transferred from callus box to nursery whilst in the tubes. The cardboard tubes, which are slightly waxed, rot away in the nursery soil.

With very ready-rooting subjects it has been possible to succeed with cuttings of rootstocks budded previously with a scion variety whilst still on the parent plant. Stems of the rose rootstock (*R. odorata*) were first budded and then, after a few weeks, taken as cuttings, cleared of stock buds but otherwise treated normally.

NURSE GRAFTING

Plants that are not readily propagated by cuttings may be temporarily nursed by grafting them to roots or cuttings of more ready-rooting subjects. A simple form of this is to inarch

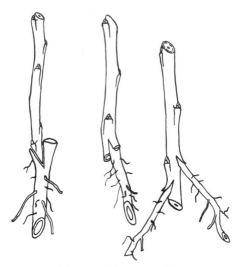

Fig. 90 Nurse grafting

Three methods of attaching nurse roots to stem cuttings. The composite cutting is planted in the normal way and, if desired, the nurse roots may be removed when the stems have established their own root system.

a piece of root into the base of a stem cutting (Fig. 90) and to plant the cutting in the normal way. When roots have developed on the shoot the plant is lifted, the nurse root cut or broken away, and the rooted shoot replanted. Variations of this technique will readily suggest themselves.

Recent developments have shown that nurse grafting has a much wider application than merely supplying temporary roots. By grafting a short piece of one-year-old seedling apple root upon the apical end of a root cutting of a very shy-rooting apple the root cutting is made to root successfully. Floor[36] nursed hardwood cuttings of apple rootstock M.9 with pieces of seedling root and obtained buddable plants in the same season (Fig. 91). Nursing can therefore be supplied by root or shoot, basally, apically or laterally. Garner and Hatcher, in the course of their vegetative propagation studies, placed scions of related and unrelated species on apical ends of stem cuttings and found a correlation with leaf-activity rather than with botanical relationship. Leafless cuttings of *Citrus* spp. which have otherwise failed to root have been

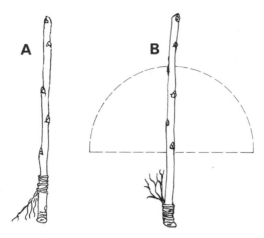

Fig. 91 Inverted seedling nurse

Apple M.9 hardwood cutting piece-root grafted with Malus seedling in winter, warm stored, planted in spring and temporarily mounded. A. Normal control. B. Inverted nurse root. B superior to A because better rooted and easier to separate at transplanting one year from grafting (after J. Floor).

induced to do so by grafting upon them leafy scions. Success here depends upon the retention of leaves by the scion.

Additional leaves. The writer found that leafy autumn cuttings of the apple rootstock Malling Crab C, side veneer grafted in their internodes with six individual leaves of the same clone, developed twice as many roots, totalling three times the length of those non-grafted. A very shy-rooting crimp-leafed genetic dwarf *Prunus avium*, given additional leaves of *P. mahaleb*, rooted vigorously. Summer cuttings of this same genetic dwarf dropped their own leaves and failed to root but rooted vigorously when nursed with individual leaves of the evergreen *Prunus laurocerasus* (Plate 16). The roots were reminiscent of those generally associated with high mycorrhizal activity. Graft incompatibility, partial or complete, does not appear to reduce nursing powers; on the contrary, complete incompatibility may be an advantage as it ensures their eventual shedding at the junction in the absence of a normal abscission layer. Leafless hardwood cuttings of the apple rootstock M.9 were each given six additional leaves

215

16. Additional leaves aid rooting

Leaves of Prunus laurocerasus *grafted on hardwood cuttings of apple M.9 in January (above) and placed under mist till March rooted better than control cuttings with no additional leaves. Softwood cuttings of a shy-rooting genetic dwarf* Prunus avium *(below) rooted heavily when given additional leaves of* P. laurocerasus *(A and B). Non-grafted cuttings (C) failed to root.*

of *P. laurocerasus* at the end of January and placed under mist (Plate 16) alongside cuttings similarly wounded, tied and sealed. When photographed in March those with additional leaves had rooted better than the controls.

Cuttings with additional leaves do well under mist. To reduce leaching during a lengthy period the upper surfaces of the leaves may be painted with a clear anti-dessicant or varnish.

Cuttings may also be induced to root by deep planting after grafting upon root pieces of a subject with which they are incompatible, the symptoms of incompatibility being delayed. Incompatibility can be assured by inverting the root piece[75]. A similar effect is obtained by binding the shy rooter with wire immediately above the graft. In due course this restricts the passage of materials and the scion puts forth roots above the wire and, later still, the nurse root drops off.

X cuttings. Shy rooting stem cuttings have been approach grafted to compatible, ready-rooting cuttings. Such composite cuttings are known as 'X cuttings' because of their shape[35] (Fig. 92). The suggestion is that the early-rooting

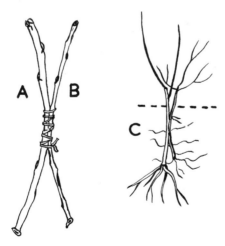

Fig. 92 Floor's 'X' cutting

A shy-rooting cutting (A) is approach grafted to a ready rooting nurse cutting (B) and separated after establishment (C).

component supplies moisture to the whole so maintaining the shy rooter until it has time to root. It has been shown that in these X cuttings it is the early roots that are important, for defoliation of the ready rooter had no adverse effect.

Suckling. The aim here is to prolong the survival of a shy-rooting stem cutting in surroundings conducive to rooting. Hardwood cuttings are planted alongside well-established plants, with which they are compatible, inarched (Fig. 93)

Fig. 93 Suckling
Hardwood cuttings nourished by an established plant until they root.

and allowed to remain for one or two seasons until they develop roots and shoots of their own, before they are separated and transplanted. Such inarched cuttings may grow shoots and fruit without forming roots; living 'backwards' as it were. Yet, in this dependent condition, they are not in rhythm with the mothering plant, for an early foliating cutting may come into full leaf, blossom and set fruit, whilst the mother is still dormant; once again demonstrating that each graft component retains its own characteristics (Plate 17).

218

E.Yoxall Jones Copyright East Malling Res. Stat.

17. Suckled but not dominated

A late-to-leaf apple rootstock (M.16) inarched with non-rooted shoots of the apple Ribston Pippin. One year after uniting, when entirely dependent on the rootstock, the new shoots have leafed, blossomed and set fruit whilst the rootstock is dormant.

Veneered cuttings. An ingenious method, of academic rather than practical interest, is to veneer graft a ring of rind of a ready rooter on to a shoot of a shy rooter. When the veneer ring is well united the shoot is taken, by cutting at the base of the ring, and treated as a cutting. Roots emerge from the ring much as they do from non-grafted cuttings of this ready-rooting variety.

BENCH GRAFTING LARGE STOCKS

When trees and shrubs are lifted for transplanting, opportunity arises for grafting under cover. Rootstocks may then be high-worked with good results. This technique is particularly valuable when certain scion varieties are scarce. For example, a fruit grower may require standard trees of a variety which is not available, except in the form of scion-wood. Trees of other varieties, used as stocks, are grafted with the desired variety in such a way that the scions form the growing points

219

of the leading shoot or shoots. Standard cherry trees, with one-year heads of three or four branches, worked some 150 mm (6 in.) from the crotch immediately before planting, have made growth equal to that of similar trees which had not been grafted.

PROPAGATING SHY-ROOTING STOCKS DURING TREE CONSTRUCTION

This is a composite technique employing bench grafting and nurse-root grafting to provide, at the same time, the benefits of vigorous shoot and root development for a shy-rooting scion thus shortening the nursery period. In autumn or spring, 40 cm shoots of the shy rooter are whip-and-tongue grafted at their upper end with the desired cultivar, tied and sealed (Fig. 94). The lower end is grafted to a whole- or piece-root of any readily available compatible rootstock. After tying and sealing, a metal strip about a centimetre wide is wound round the base of the interstem immediately above the nurse-

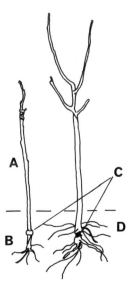

Fig. 94 Propagating shy-rooting stocks during tree construction
A. Shy rooter bench grafted with desired cultivar. B. Nurse root bench grafted to base of shy rooter. C. Metal band locked round base of shy rooter. D. Scion roots above constricting band (refer to text for full description).

root. This strip must be locked or otherwise fixed to ensure eventual constriction. At planting, the lower union should be buried 15 cm so that the interstem can scion-root. After one season the tree may be transplanted but if weakly it is better left a further year. If adequately scion rooted, the redundant nurse-root may be broken off but, otherwise, it is left to continue its nursing function, being discarded later by abstriction.

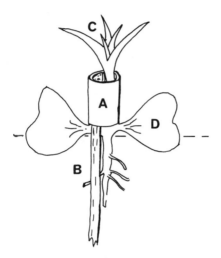

Fig. 95 Tube-and-splint support

A. 12 mm (½ in.) length of thin plastic tube holding splint and graft. B. 75 mm (3 in.) bamboo splint inserted in rooting medium. C. Scion, wedged or spliced to cotyledon-stage seedling. D. Stock with active leaves.

CHAPTER VII

Tree-raising in Nurseries

The large-scale propagation of plants by grafting is usually left to specialist nurserymen, and with rare exceptions this appears to be the most satisfactory arrangement. An established nursery has trees at various stages in production and can, at reasonable notice, supply the more common varieties in quantity. To meet the demand for his product the nurseryman organizes his business rather like a factory and simplifies his technique to the utmost. It follows that he will not use complicated methods when the simple ones are adequate, and therefore, except for a very few subjects, he confines his grafting methods to less than half a dozen. It will only be necessary here to give a few examples and to refer the reader to Chapter VI for the remainder.

It is convenient to divide tree-raising in nurseries into two classes: in the open, and under glass, plastic or other protection.

TREE-RAISING IN THE OPEN

Fruit Trees and Kindred Subjects

Trees of the hardy fruits are almost invariably propagated in the open. Many hardy ornamentals are also similarly treated and a description of one will also serve for the other.

The rootstocks, propagated as described in Chapter III, are planted in rows spaced conveniently for cultivation. The actual spacing depends upon the size of the young tree and the number of seasons it is to stay in that position. Thus, vigorous growing subjects such as plums and sweet cherries should be spaced at about 380 mm (15 in.) in rows 1·2 m (4 ft.) apart, whilst apples and pears, especially where they are to be transplanted as maidens, may be set 300 mm (12 in.) apart in rows 0·9 m (3 ft.) or more apart. The age at which the tree is to leave the nursery should be considered, and trees which are to remain in the nursery for long periods should be assigned a drift apart from the others, so that each drift may be cleared at one time.

Land that has borne agricultural crops for some years yields better results than land recently under nursery cultivation, and wherever possible 'fresh' land should be used for tree-raising. Newly-broken turf usually induces over-vigorous growth, which may quickly succumb to disease, and cannot be generally recommended. The layout of the nursery should be carefully planned and the rootstocks planted in straight lines. This is known as lining-out. Some nurserymen always plant rose rootstocks in rows running from east to west and they lay the rootstocks over to the south and insert the buds on the north side. This is said to protect the bud from the worst effects of alternate freezing and thawing during cold weather.

There are various methods of planting rootstocks. In all cases the nursery should be free of perennial weeds and deeply cultivated. One of the common methods is to plough or dig the soil and to fix the position of rows by marking against a line. The line is removed and the rootstock planted firmly in a grip or hole made with a spade, the base of the rootstock being 150–175 mm (6–7 in.) below the soil surface.

Tractor-drawn machines, designed to plant cabbages, celery and similar plants, have been modified to plant rootstocks. Soil conditions must be really good if the material is to be well planted, and the machine must travel slowly to give time for placing the somewhat variable material at a suitable and regular spacing. Well-designed machines can do excellent work. If the rootstocks have not been shortened before planting they are now cut down to about 450 mm (18 in.).

The newly-planted rootstocks are encouraged to make good growth by complete weed control during spring and summer. The growth of young trees in the nursery is largely decided by the growth of the rootstock before it is budded or grafted. Application of stimulating manures after planting does not compensate for lack of early planting and weed control.

The rootstock stem should be kept clear of sprouts for a sufficient distance above ground so that there may be room to insert the bud or graft unhampered by side-growths. Sprouts should be removed, while soft and green, by hand rubbing.

METHODS OF GRAFTING

Only three methods need be used by raisers of fruit trees and most of the ornamentals, namely shield budding (page 144), chip or plate budding (page 158), and whip-and-tongue grafting (page 176). The budding is done in the first summer after planting, and grafting with dormant scions is done in the following spring. Other methods are also used for particular purposes as already described in Chapter VI. Stages in the normal procedure are illustrated (Fig. 96).

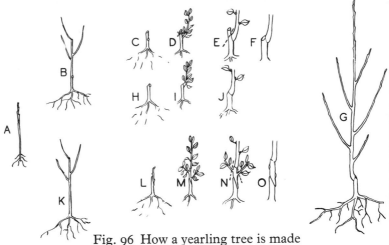

Fig. 96 How a yearling tree is made

A. The newly planted rootstock. B. Rootstock budded in summer and dormant in winter. C. Rootstock cut down leaving a 100 mm (4 in.) snag above the bud. D. Bud grown out and shoot tied loosely to snag. E. Snag removed in early autumn. F. 'Snagged' dormant tree. G. Resulting maiden tree.

H. Rootstock cut down close to bud (alternative to C). I. Bud grown out. J. Tree in autumn, no snag to be removed. G. Resulting maiden tree.

K. Established non-budded rootstock dormant in winter. L. Whip-and-tongue grafted in March. M. Growths from scion. One selected and two pinched at five leaves. N. Region of union trimmed in early autumn. O. Dormant tree. G. Resulting maiden tree.

Budding. Budding is done when the rinds readily part from the wood of the stock and when buds have developed at the base of the leaf-stalk on young shoots of the scion variety. This means that the budding season may extend from early June until September, but from the end of June until mid-August is the optimum time in normal seasons.

In periods of prolonged drought the rind may not lift without tearing, and it is then wise to postpone budding in the hope that rain will come and the rootstocks increase their rate of growth. In the absence of rain or adequate irrigation, the rootstock should be grafted the following spring.

The collection and care of the scion-wood, or bud sticks, has been described in Chapter IV and the actual operation of shield budding in Chapter VI.

Many fruit trees, particularly apples, are worked on to dwarfing rootstocks close to the ground, and there is every likelihood of the scion putting forth its own roots[62], should it come in contact with the soil. Buds should be inserted into rootstocks at least 150 mm (6 in.) above ground level in order to facilitate planting the tree with the union above ground and so avoid the risk of scion-rooting (Plate 18). Apple growers, using dwarfing rootstocks, prefer to plant really large maiden trees already possessing suitably-spaced laterals and balanced root systems. Such trees are only obtained under best nursery conditions and good management[68]. The rootstock must be virus-free, large, and well rooted, and be planted early on fresh, or fumigated, land kept free of weeds. The virus-free scions should be placed 300–450 mm (12–18 in.) above ground level so that the well-angled primary laterals develop where needed.

Inserting buds on any particular side relative to compass bearings has not generally proved of much practical benefit except where the bud position is exposed to intense sunshine. As far as possible the buds should be inserted on the same side of the rootstocks in a nursery, so that they may be readily inspected and manipulated subsequently. The operator proceeding along a row will find it convenient to insert the buds all facing in the same direction.

Some rootstocks have bushy heads which obscure a view of

18. Scion-rooting

A. A Doyenné du Comice pear on a satisfactory quince rootstock (Malling A). B. The same variety on an unsatisfactory quince rootstock, planted at the same time and cultivated similarly. C. The union between the pear scion and quince rootstock inadvertently covered with soil, into which the scion has rooted, resulting in a very vigorous, upright and unfruitful tree. D. Roots of an older scion-rooted pear tree. The original quince roots (painted white) have been superseded by very vigorous pear roots from the scion.

the main stem, but the propagator can overcome this difficulty by placing his legs a few inches either side of the rootstock so that the branches are behind his knees and the bud is inserted in the rootstock between his ankles. The actual insertion of the bud is described on page 144, and methods of tying on page 148.

Some operators make a habit of licking prepared scions in order to moisten them or to lubricate their entry into the stock. Experienced grafters have observed that the wetting of scions in this way is no advantage and the practice is generally condemned. Dr. T. Raptopoulos of the University of Athens informed the writer that one budder's lack of normal success in budding apricots on to almond rootstocks in Greece was definitely traced to excessive wetting of the buds with saliva, presumably with a view to lubricating their entry. When the operator was weaned from this habit his successes rose to normal. It appears that any profuse wetting of the cut surface of scion or stock may prove detrimental.

The abscission of the leaf-stalk is one of the signs of success. Shrivelled adhering leaf-stalks, not held by raffia, indicate failures and, if the season is not too advanced, another bud may be inserted in a fresh place. Persistent low take of buds has sometimes been overcome by smearing buds and all incisions with petroleum jelly immediately after tying. This treatment also prevents damage by the red bud borer (page 120). Other than the incisions necessary for the insertion of the buds, no slits should be made in the rootstock rind at this season as these wounds provide an entry for bud borer which, in a severe attack, may destroy the cambium all around the stem, causing the death of both bud and rootstock. After budding, the soil should be kept free of weeds to encourage further growth. Buds normally remain dormant until the following spring. If growth occurs before the autumn, it may be cut back in the spring to one or two visible buds at its base.

In January or February the buds are inspected and failures marked by cutting the head of the rootstocks higher than where the buds are successful so that the grafters find them readily. The rootstocks with successful buds are usually cut down to 100 mm (4 in.) above the bud; this section, known as a 'snag', serves as a stake to which the bud growth is tied. Alternatively the rootstocks are cut off immediately above the bud, the growth from which is not supported in any way. Such treatment saves labour and permits the healing of the wound during the first season. In both cases the wound should be sealed. Where a 'snag' is left, it is removed towards the end of the summer following budding. A certain amount of skill is necessary for this operation to be neatly accomplished.

Sucker shoots will arise on the rootstocks during the spring and early summer and these must be removed on emergence, either by rubbing or slicing off with a knife. Where a 'snag' is left, the growth from the bud is tied to it when it is 100–125 mm (4–5 in.) long and still supple. Loose tying is necessary to permit the swelling of the growth without constriction. It is now common practice to permit all laterals to grow unchecked during the maiden year. Such laterals are known as feathers, and one-year-old trees furnished with them as feathered maidens.

Whip-and-tongue grafting. The failure of buds is normally made good by grafting with dormant scions in March. Non-budded rootstocks are also grafted at this time. In both cases it is usual to cut off the head of the stock at the time buds are inspected in January or February.

When about to begin grafting, cut the scion-wood into selected lengths of three or four buds, not more, in sufficient quantity to last for one or two hours' work. The slender unripened tips of the shoots are not used. The upper end of each piece should be cut close above a bud so that it will heal without a dead snag; apical cuts should always be made in this way. In normal tree-raising the length of the whip-and-tongue scion is governed by the length which can be comfortably held when slicing and tonguing: 100–125 mm (4–5 in.). The number of buds thereon is immaterial, provided at least one is viable, though it is good to have one or two in reserve. When the material is scarce it may be necessary to use single bud lengths and to slice and tongue before separation from the shoot, because the very short piece cannot be gripped whilst cutting. The tongued piece may then be cut into a box, or onto a cloth, without becoming dirty. A shallow box with a large handle should be provided to carry the selected scion-wood. Details of the preparation of scion and stock may be found in Chapter VI.

In sealing the graft with wax, all cut surfaces, including the apical end of the scion, and the ties over the part of the scion in contact with the stock, should be well covered, but not the ties over the back of the stock, which should be left exposed to permit of easy cutting when constriction is likely to occur at the end of May or early June. Some adhesive tapes (page 120) often burst without constriction, but care is necessary in case this does not occur. Sealing over the tape is not necessary but any exposed, cut surfaces are best covered with wax or grease.

New growth from both scion and rootstock appears towards the end of April or early May. That from the rootstock should be removed before it is more than 25 mm (1 in.) long, but all scion growths should be left untouched until the longest is 150–200 mm (6–8 in.) in length, normally towards the end of May or early June. At this time the best shoot, usually the apical, is selected to form the trunk of the future tree, and the growing tips of the remainder are pinched out, leaving five or six leaves on each, to aid the growth and formation of a sound union, until the autumn when these pinched shoots are cut clean away. A sharp knife is now passed up the back of the rootstock to cut the ties without injuring the bark. The severed ties should not be unwound for fear of disturbing the union.

Staking is not absolutely necessary for all types of scion varieties, but it is customary and even vital with some subjects and in exposed situations. Newly grafted apples and pears may be efficiently secured by the use of a cane, wand or packing-stick, no more than 600 mm (2 ft.) in length, pushed into the soil 150 mm (6 in.) deep, close to the rootstock. The rootstock is tied to this below the union and the scion growth is looped to the support about 150 mm (6 in.) above the union. Varieties of plum that grow extremely rapidly and remain slender are tied, as they grow, to 1·5–1·8 m (5–6 ft.) canes held in line by a wire at 1·4 m (4 ft. 6 in.) above the ground and attached to stakes at frequent intervals. Rigid support by the use of thick stakes not only prevents the formation of taper but by unilateral shading causes the tree to grow away from the vertical, involving more frequent tying. With many varieties of apple and pear staking is unnecessary, but the risk of breakage may be lessened by placing a single knotted tie towards the upper part of the union when the original ties are cut. This single tie is severed some three or four weeks later, when a really strong union has been made. All laterals on the selected growth, except occasional ones that may bifurcate the tree, are left intact during the maiden year, as this causes stout growth of the stem towards its base. In late August or early September, at the time the 'snags' are removed from budded maidens, the grafted trees should be trimmed at the union by removing any small unhealed portions of the rootstock and the bases of the unwanted growths from the scion, which were stopped when showing five leaves in the early summer.

SPEED IN GRAFTING
In budding the adept has the best chance of success, because the moist tissues exposed in the operation have little time to dry, but excessive speeds involving hasty work are no advantage and 'speed records' are not to be encouraged. The novice should endeavour to master the technique before attempting rapid work. Obviously a 10 per cent increase in speed is harmful when it involves a 1 per cent decrease in 'take', for the individual operation is a matter of seconds, but the young tree must stand under nursery cultivation for two or three years. Nevertheless, an indication of normal safe speeds may be of interest to the beginner. With a budder and tyer working together, with the prepared scion-wood at hand, one hundred buddings per hour is a good average speed that can be maintained throughout the day by practised workers. Equally-skilled workers could put on some seventy whip-and-tongue grafts per hour, inclusive of waxing.

HIGH-WORKED TREES
Trees are sometimes grafted where the head is formed. Standard sweet cherries are usually high-worked. Other trees may be stem-built by grafting the rootstock near the ground with a vigorous clean-growing variety, which is later high-

worked with the kind desired. In simple high-working good results are obtained by lining-out rootstocks, as described earlier, cutting them down to 150 mm (6 in.) *twelve months after planting*, selecting one good shoot from this stub to form the trunk and removing all others on appearance. When this trunk has passed the height intended for working, it is grafted just below the position of the future branches and these are taken direct from the scion. Four- or five-bud scions should be used, and they will usually provide three good branches. Alternatively the trunk may be budded with one or more buds at the desired height, as previously outlined. Manipulation is more difficult at 1·8 m (6 ft.), but the trunk or stem is usually supple and may be temporarily bent over and held under the grafter's arm.

A number of cherry trees and other subjects are high-worked by grafting in late September or early October with defoliated scions. The whip-and-tongue method is used and the graft is carefully sealed. The following spring the scions make a good start and are not jeopardized by adverse spring weather.

HIGH-WORKED BUSH FRUITS

In Germany gooseberries are commonly worked high on stems of *Ribes aureum* both for private garden use and for fruiting in hedge form. A method has been developed which involves side grafting in the open in August. The *R. aureum* rootstocks are raised by stooling (*see* page 90) and the rooted layers are cleared of all buds from the root-bearing area, and from the lower 300 mm (1 ft.) of stem, to check future suckering, otherwise the rootstocks are neither trimmed of laterals nor shortened. These prepared rootstocks are lined out in the nursery and are tied by a withy to a single wire placed about 150 mm (6 in.) above the position of the future graft. The grafting is best done in the August following lining-out, when the scion shoots have become firm and have formed terminal buds. A well-established, lightly-pruned bush provides suitable shoots for scions. The aim is to have well-ripened scions while the rootstocks are in active growth; normally this situation does not last more than a week or so. The gooseberry scions are cleared of leaves and thorns and

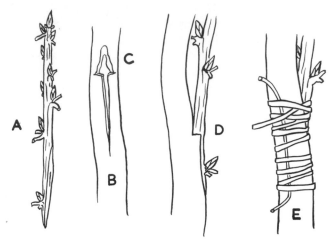

Fig. 97 Side rind graft
Used for high-working bush fruits.

prepared with one slicing cut about 50 mm (2 in.) long (Fig. 97, A). The rootstock is cleared of laterals in the vicinity of the proposed graft (B) and cut as for normal shield budding, save that a wedge of rind is removed from above the transverse cut (C). The scion is pushed down under the lifted rind (D) and the graft is firmly bound. To avoid risk of cutting the stock rind when removing raffia ties, a piece of thin galvanized wire is placed at the rear of the stock before tying (E). The graft must be well sealed. In the following spring the tie is cleared by pulling the wire. At this time, when the risk of severe frost has passed, the stock is cut through a few inches above the support, to which it remains attached, and all stock laterals and buds are removed. When the new scion shoots are long enough, but still soft and green, they are loosely bunched up to the snag so that the bases of the future branches are turned upwards. The following autumn, some fourteen months after grafting, the snag is removed and the wound carefully waxed prior to transplanting.

If the seasonal conditions at grafting time prevent the use of the rind graft method, the spliced side graft (Fig. 79) is the alternative. Should both these fail the stock is beheaded a few inches below the failed graft in spring and regrafted with stout dormant scions by the inlay method (Fig. 48). A withy, or string, is tied below the graft, and then on the wire above, to serve as a support for scion shoots.

DOUBLE-WORKING

A double-worked tree has an intermediate stem piece of another variety between the upper scion and the absorbing root system. This intermediate piece may be a mere sliver, as short as a few inches or as long as the whole length of the stem or trunk and in some cases may extend to include the bases of main branches immediately above the crotch. Thus double-working has various purposes which may be considered conveniently under five heads; (1) to overcome incompatibility between scion and rootstock; (2) to build trees with strong, straight stems (stem-building); (3) to increase resistance to disease by providing resistant stems; (4) to promote frost hardiness; (5) to curtail vigour and to increase fruiting.

To overcome incompatibility. Double-working to overcome incompatibility is comparatively simple and inexpensive when it is done in one grafting operation as described here (Fig. 98). Dormant scion-wood of both intermediate and upper scion is collected and stored in the usual way as described in Chapter IV. The scion-wood of the intermediate variety, selected for its compatibility with both rootstock and scion, is cut into 125 mm (5 in.) lengths in March and immediately grafted with a three-bud scion of the chosen upper scion variety. The completed grafts are sealed and stored upright in a cool, moist situation. Damp peat is suitable, but only the lower half of the amalgamated scions should be buried. If gritty substances such as sand or soil are used they may blunt the knife at the second grafting and should be washed away upon removal from storage.

The intermediate, with its scion attached, is subsequently grafted on to the established rootstock in nursery rows. If desired the first graft may be made some weeks before the second. When complete, the length of the intermediate between the upper lip of the stock and the lower lip of the upper scion should not be more than 50–75 mm (2–3 in.).

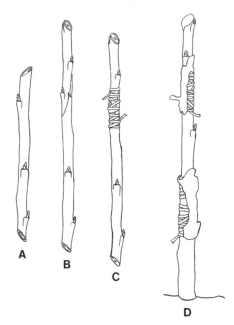

Fig. 98 Double-working in the nursery

A. A 125 mm (5 in.) length of intermediate (Beurré Hardy). B. A three-bud scion of Williams' Bon Chrétien grafted to the intermediate of Beurré Hardy. C. Tied and ready for sealing before setting aside. D. The worked scion joined to the established quince rootstock in the nursery and sealed with grafting-wax.

Provided that the rootstocks have become well established during the previous year, the maiden, double-worked trees will reach the same height and girth as well-grown, single-worked trees of the same variety, and similar stands will be obtained.

A still simpler method of double-working in one operation is used when budding. This has worked well for pears that are unreliable when single-worked on quince rootstocks[43] notably the pear Williams' Bon Chrétien (or Bartlett). When shield-budding with a T incision, a small budless shield of an intermediate variety, e.g. Beurré Hardy or Fertility, is slipped into the incision (Fig. 99), a bud of the upper scion variety is placed on the budless shield and both are slid home together.

234

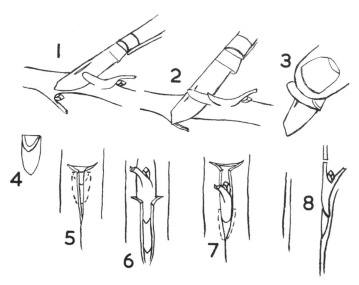

Fig. 99 Double-shield budding

1. A shallow cut in a shoot of intermediate. 2. A second cut to procure the budless shield. 3. Shield held between knife and thumb. 4. Outer view of prepared shield. 5. Dotted line shows position of shield in the rootstock. 6. Bud shield on top of budless shield before sliding both down into place. 7. Shields home. 8. Diagrammatic side view.

The budless shield is quite small when inserted but grows rapidly and eventually extends around the tree to form a reliable bridge between the quince rootstock and fruiting pear variety. Double-shield budding compares favourably with double grafting with dormant scions, not only in being easier to do, but also in economy of material, for many budless shields may be cut from a single shoot of the intermediate variety.

It is obviously important to be able to recognize successfully double-worked trees, even though many hundreds of double-shield unions have been examined without discovering any failure of the intermediate. There are two distinctive features: (1) the rootstock stem thickens more below the maiden growth than below the snag wound on the opposite side, whereas the rootstock below single bud unions thickens almost equally on all sides but fails to grow as much as the pear above, (2) the snag wound in double-shield trees remains uncovered for some years, but in single-worked, a roll of callus soon begins to cover it. Another double-shield method is described in Chapter VI, page 149.

There are other methods of double-working to overcome incompatibility, the most common of which requires an extra year to produce a maiden. The rootstock is budded or grafted with the chosen intermediate, as in single-working, and one year later this maiden tree is cut down to within a few inches of the union and grafted with the upper scion, which then needs a year in which to reach the maiden or whip stage. A modification of this, which is also useful in reworking maidens of unwanted varieties, is to graft the maiden in the position desired for the head of the tree with a four-bud scion forming the head of the tree direct from the scion, so saving the additional year required by the previous method, in which the maiden is cut down close to the first union*.

Stem-building. Some scion varieties are slow to form stems or habitually grow away from the vertical and so prove difficult to grow into first-class standard or even half-standard trees. The use of suitable stem-builders for these varieties provides trees with straight vigorous stems in the minimum time.

Stem-built trees are produced by working the rootstock in the normal way with the chosen stem-builder and working this intermediate variety at the place where the head of the tree is to be formed. When the intermediate is required to form the whole of the trunk of the tree, as it usually does in stem-built standard trees, it may be necessary to permit the development of feathers upon the intermediate to encourage sturdy growth. In such trees the roots will be influenced for a time by leaves of both intermediate and top scion variety and normal growth may not occur, but so little is known of these complicated interactions of rootstock, intermediate and top scion that, for the present at least, they are ignored in commercial practice.

Disease-resistant stems or frameworks. Some otherwise desirable scion varieties have proved to be susceptible to diseases which attack the trunk and limbs of the tree. In these cases it

---

* See Appendix I for list of pears improved by double-working.

is often an advantage to form the trunk and bases of main branches of a variety which is more resistant to these diseases. Recent research has proved that it is possible to select definitely disease-resistant varieties as stem-builders for the production of trees of susceptible varieties. Where such resistance is present in the rootstock, there is no necessity to work the rootstock at ground level, but it should be run up to the required height prior to working. For working above the crotch the stem is cut to form a head of three or four branches, which are individually grafted when one year old, such foundations are commonly known as staddles, an ancient term for a base, foundation or support.

Where stems or trunks of trees are liable to attack by collar rots of various kinds it is desirable to use resistant varieties at vulnerable points. To increase resistance of apple trees to collar rot, caused by *Phytophthora cactorum*, it is merely necessary to work them on a resistant rootstock about 300 mm (1 ft.) above soil level.

An unusual use of stem-building is reported from Ceylon[134]. Rubber tree seedlings are worked at ground level with a clonal selection which, though high-yielding, is susceptible to mildew. This stem is later high-worked with a resistant clone which forms the head of the tree. Though this low-yielding top scion may have a depressing effect on the yield of the intermediate, this is limited to the upper part of the trunk and can be avoided by having a trunk of intermediate long enough to provide a tapping panel of 0·9–1·2 m (3–4 ft.) in the lower, high-yielding part of the tree.

To promote frost hardiness. In countries where severe winter frosts may injure or kill fruit trees it has been found[33] that the incorporation of frost-hardy varieties in trees of tender varieties confers a high degree of resistance. In Poland the hardy variety of apple Antonowka is used to form the trunk and main branches, which are later worked with the desired fruiting variety. It has been found necessary to retain some unworked branches of the hardy variety on the most frost-susceptible S.W. side of the tree. These branches must not be swamped by branches of the tender variety or their good effect will be lost.

To curtail vigour and increase fruiting. The simple and proved method is to employ the appropriate rootstock. In areas where suitable rootstocks are not yet available use has been made of double-working to incorporate intermediates calculated to simulate rootstock effects. One method is to bench graft the available rootstock with a dwarfing scion and when this is established in the nursery to bud it with the desired fruiting variety. In countries where bench grafting is not customary, double-working in one operation (*see* page 233) is appropriate. It would seem that the effect of a given intermediate is in the same direction as a rootstock of that variety though not always so definite. An intermediate of quince in a pear tree markedly curtails vigour and induces early blossoming as shown in Plate 19.

In some areas it has been found that dwarfing rootstocks are insufficiently anchored against high winds and it has become customary to use vigorous rootstocks of good anchorage and intermediates (interstocks) of a dwarfing nature to curtail vigour and encourage heavy cropping. An example of this is seen in the U.S.A. where seedling apple rootstocks are worked with a 'Clark Dwarf' intermediate of 75–100 mm (3–4 in.), and then with the fruiting variety.

PLANT-LIFTING MACHINES
The lifting of individual specimens or small groups in the nursery is best done by hand, but where whole rows are cleared at one time lifting machines can do the work more rapidly and economically. There are many kinds of lifters, from quite small ploughs (Fig. 100) to the huge machines used in very large nurseries. In between these there are many useful patterns. One of the best for bushes and young trees designed by Mr. Eric Chudleigh of Dixie, Ontario, and adapted for use in England by Mr. Denys Hammond of Maidstone, is of very simple construction (Fig. 101). This implement is attached to the hydraulic lift of a high-clearance tractor and is handled by three men inclusive of the driver. The trees, loosened in the soil, may be drawn immediately, or later, as required.

Eleanor C. Thompson

Copyright East Malling Res. Stat.

19. An effect of the intermediate

Double-worked pear trees of equal age grown side by side. The combination Conference/Conference/Pear Malling C8 (left), gave vigorous shoot growth, coarse vertical roots and delayed blossom, whilst Conference/Malling Quince A/Pear Malling C8 (right), resulted in weaker growth, fibrous, spreading roots and profuse blossoming.

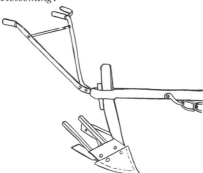

Fig. 100 Plough for lifting lined-out shrubs, rootstocks and small trees

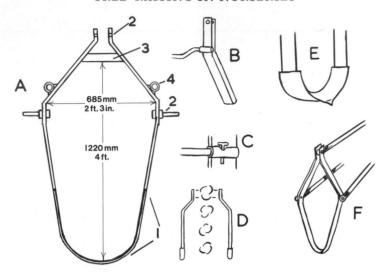

Fig. 101 Machine for lifting bush fruits and small trees
A. Frame of 100 × 12 mm (4 × ½ in.) steel. 1. Sharpened leading edge. 2. Positions for tractor links. 3. Welded tube to strengthen. 4. Handle sockets of piping. B. Side view showing forward cant of cutting leading edge and position of handles. C. Handle pegged in pipe socket. D. Position of handles in relation to trees. E. Lifting share as alternative to sharpened frame. F. Three point attachment to hydraulic lift of tractor.

GRAFTING UNDER GLASS

Many subjects are grafted under glass both in the herbaceous and the woody state. Plants that are grafted only with difficulty in the open often respond favourably to treatment under controlled conditions. The most common method of grafting green-wood or herbaceous plants is by the simple wedge (page 186). Gentle tying is essential to avoid crushing the soft tissues. It is helpful to loop the ligature loosely around the stock, insert the scion, raise the ligature and tighten it round the graft. In many grafts, done under glass, the sealing of cut surfaces is not necessary, though some operators insist that it is beneficial.

The grafting of the walnut by the big-tongue graft under glass has already been described (page 178). The techniques used in grafting roses, rhododendrons and azaleas, sorbus

240

and spruce, now to be described, are applicable to many other woody subjects. The herbaceous grafting of the double gypsophila on to *G. paniculata* and the splice grafting (page 245) of brooms (*Cytisus* spp.) on laburnum (*L. vulgare*) provide other examples of grafting under glass.

Roses

The production of roses under glass for the cut-flower trade is a highly specialized industry. The bushes are propagated by grafting the young rootstocks, usually *Rosa canina* seedlings but also *R. manetti*, from November to May. The earlier grafting makes use of dormant or 'pushing' scion-wood and is known as winter grafting. The later work, with current season's scion-wood, is termed spring grafting. The rootstocks are potted into 90 mm ($3\frac{1}{2}$ in.) pots in October. The best size is 5–7 mm ($\frac{3}{16}$–$\frac{7}{32}$ in.) at the collar but larger stocks can be used. Immediately after potting they are plunged in an ash-bed in the open to become root-established. A fortnight before grafting, the potted rootstocks are taken into the propagating house where they are placed on an open bench, air temperature 16–18°C (60–65°F).

For winter grafting the scion-wood is collected from dormant, established bushes under glass at the time of grafting. Each scion includes some 50 mm (2 in.) of the base of the most recent growth and a similar length of the previous growth (Fig. 102, A). Such scions give rise to bushes with well-furnished bases. Straight scions, entirely from the most recent growth, do not furnish so well. For spring grafting the bases of the flowering shoots are collected for scions. The leaves are cut from the scions immediately after collection. The graft used is the simple splice with a short cut surface (B). Stocks considerably thicker than the scions are rind-crown grafted (C) but, wherever suitable, the splice graft is preferred. Both grafts are tied with raffia and sealed with painter's knotting (a thick varnish). This sealing is not universally practised, for some propagators leave the grafts quite unsealed and yet have excellent results in the humid atmosphere of a close propagating case.

The completed grafts are placed in a case at 21–22°C (70–72°F). Growth begins almost immediately and when the

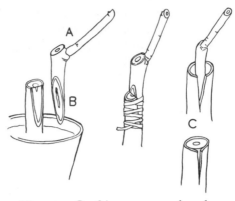

Fig. 102 Grafting roses under glass

The more popular splice graft and the rind method used when the rootstock is much larger than the scion. Both methods must be tied but sealing is not essential.

young sprouts from the scion are 25 mm (1 in.) in length, usually seven to ten days from grafting, air is gradually admitted and, after a further ten days, the plants are moved to the open bench. Three or four weeks later still, they are planted out in their permanent places in the forcing houses. At every stage the weakly plants are eliminated and it is essential to graft considerably more plants than will finally be required.

The supplies of bushes for forcing may be augmented by budded plants from open ground which are planted in the autumn or winter, but grafting under glass is the chief mode of propagating roses for forcing.

When new varieties of rose are first introduced, an expert nurseryman will buy what he can of the stock, and use all the wood as single-bud scions for splice grafting. The young bushes are planted out under glass to provide buds for shield budding. By this means large supplies of the new variety are quickly obtained.

Rhododendrons and Azaleas

Two- and three-year-old seedlings of *R. ponticum* or, less commonly, *R. catawbiense*, are established one year in pots and grafted with selected varieties of rhododendron during the period January–March or, to a lesser extent, in August.

Rhododendrons are grafted as low as possible on to their rootstocks so that the union may be well buried and rooting from the scion can occur whilst opportunities of suckering from the rootstock are diminished (*see* page 210).

In Belgium and elsewhere propagation of the azalea (*A. indica* of the trade) is a highly-specialized industry carried out on a large scale for export. The rootstocks generally used are rooted cuttings of related species. The rootstock cuttings are collected from one- and two-year-old plants between October and March and are struck in a medium of half leaf-soil and half sharp sand in closed cases. The strongest rooted cuttings may be grafted in May, when 150 mm (6 in.) high; the more backward in June or in the following year. The rootstocks are not stopped (pinched) but run on until grafted.

The methods of grafting for both rhododendrons and azaleas are the saddle (page 183), the wedge (page 186), or the veneer side graft (spliced side graft, page 201).

The wedge graft as used for azaleas is known as the *demi-fente en tête* (half-cleft on top). The rootstocks are cut transversely, 100 mm (4 in.) above the soil in the pot, and cleft radially (Fig. 103, A). A sap-drawer leaf is left near the top of the stock, which is now half-cleft (B). The scion (C), made into a sideways wedge, is inserted and tied with cotton thread (D). The grafted plants are placed in closed cases, temperature about 16°C (60°F), where they are inclined with

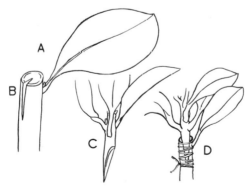

Fig. 103 Wedge grafting the azalea
Note the sap-drawer leaf on the rootstock.

the scion uppermost and the leaves touching the glass. After three weeks air is admitted, and in seven the lights are entirely removed and the scion-growth is pinched at four or five leaves to form bushy plants. The plants are plunged outside until the autumn, when they are returned to the houses for protection against frost. The majority of the plants are sold when two or three years old.

Some consider it advisable to reduce the leaf area on the scions at the time of grafting to reduce water loss. Removal of half of each leaf affords ample reduction. Live sphagnum moss placed around the grafts in the case helps to keep a moist, but healthy, atmosphere until the unions are complete.

In order to make full use of propagating cases, staging or temporary benches are erected in the propagating house, or in an adjacent structure, so that when able to stand exposure to full light and air of the open house, the grafted plants may leave the cases for the open bench, and these cases may be occupied by another batch of grafts.

Sorbus (*Aucuparia* Section)

Two-year-old transplanted seedlings of *S. aucuparia* are brought under cover one or two days before grafting which may begin in mid-February and continue through March. Early grafting is best but later grafting, with cold-stored scions, can be quite successful. The stocks are trimmed and the grafting area is wiped free of grit. The scion-wood, collected in midwinter, of well-developed previous season's growth, is brought in at grafting time. The stocks are splice grafted in the region of the hypocotyl, tied firmly with rubber strip and the complete scion and union are dipped in paraffin wax or other adequate seal. The grafted plants are placed in boxes of peat for ease of handling, with their unions just covered. The boxes are held in glass or plastic structures providing a bottom heat of around 16°C (60°F). Hardening off begins in about three weeks, followed in due course by containerization. Careful handling during the hardening and 'growing on' period is essential. This includes adequate compost nutrition, shading and watering for some two months from grafting.

Spruce (*Picea* spp.)

The popular 'Blue Spruce', *P. pungens* '*Kosteriana*', the 'Koster' spruce, will serve as an example. It is grafted on seedlings of Norway spruce (*P. abies*). These seedlings, after two years in the seed bed and one year in 75 mm (3 in.) pots, are brought under cover in early winter and kept fairly dry. At no time must the plant be wet; otherwise the graft will be flooded by rising sap (bleeding). Grafting takes place January to March inclusive or, alternatively, in August or early September. Fourteen days before grafting the potted root-stocks are brought into a warm glasshouse at 10°C (50°F). They are veneer side grafted (Fig. 79). The scions should have a prominent terminal bud and at least three radial buds. The needles in the grafting area are very carefully removed to avoid resin ooze. Likewise the stock is cleaned up without wounding. The knife blade is kept free of resin by wiping on a pad soaked in white spirit or turpentine. Very firm tying is obtained with waxed cotton string or strong rubber strip. The grafted plants are plunged in peat in a closed case at 18°C (65°F), sloped at 45°, with the scion uppermost, the union just buried in the peat. The somewhat dry régime must be maintained. The stock is headed back in three stages, beginning six to eight weeks after grafting; then when the apical bud swells, and lastly, when the plant is completely hardened off and ready for planting out, some four months after grafting, at which time the graft tie should be released.

Gypsophila

Stock plants used as a source of scions are forwarded in heat in late winter, and when the new shoots are 75–100 mm (3–4 in.) long they are taken for scions. The root of a seedling plant is used for the stocks after all stem tissue has been removed by severing about 25 mm (1 in.) below the junction of stem and root. The top of the stock is split from one side for about 25 mm (1 in.) down. Thin shavings of rind are removed from both sides of the base of the scion, which is then inserted into the cleft in the stock, and the graft is tied, potted into a small pot, and placed in a closed case until union is complete, when the plants are gradually hardened off for planting outside.

Broom

Varieties of broom are splice grafted upon laburnum roots. Whole or piece roots of laburnum about the thickness of a pencil, containing no stem tissue, are severed transversely (Fig. 104, A) and spliced at the side with the prepared scion of broom. The scions are trimmed (B) before grafting. This grafting is done at a bench in late winter or early spring and the completed grafts are immediately potted and plunged in closed cases until established.

An inlay graft is also commonly used in working broom on laburnum (Fig. 104, C and D).

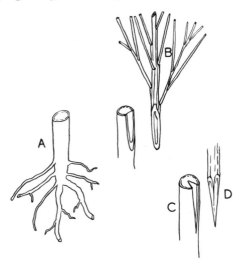

Fig. 104 Broom bench grafted on laburnum
A and B. The splice method. C and D. The inlay method. Both need firm tying.

CHAPTER VIII

Grafting Established Trees

When trees have left the nursery and have become established in their permanent positions, it is sometimes necessary to make use of grafting to change the variety or to repair and invigorate the growth. The methods of grafting used in the nursery may be used in the rehabilitation of older trees, but certain peculiar techniques have proved particularly suitable and, for ready reference, are brought together for description in this chapter.

Whilst it is true that all trees, except possibly the palms and their allies, may be successfully grafted in one way or another, fruit and other plantation trees are more frequently grafted than others. The methods which have proved so successful with deciduous fruits may be adapted to other subjects.

Whenever it becomes necessary for a fruit grower to change the variety of a tree he may do so in two main ways. He can remove the tree and plant another or he can graft new scions upon the existing tree. Grafting, properly done, is preferable to replanting except where the trees are extremely decrepit.

Grafting is essential for the preservation of trees girdled by livestock or for those severely damaged by implements. The root system as well as the branch system may be changed by grafting, and weakly trees invigorated by inarching them with more vigorous root systems. Thus there are many opportunities for the grafter in the improvement of established trees. This is considered under two main headings: to change varieties, and to repair or improve.

TO CHANGE VARIETIES

Apple, pear, and other trees have a productive life of fifty years or more, hence, when a variety becomes obsolete, the grower is loth to discard his trees at twenty years of age to plant afresh and he looks with favour on the sounder methods of regrafting (reworking). These methods are conveniently placed in two groups: (1) topworking, in which the branches are removed almost entirely and the cut ends set with scions of the desired variety, or (2) frameworking, in which the main framework branches are entirely retained and furnished with new fruiting laterals.

In both topworking and frameworking many different

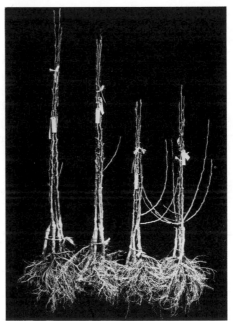

Copyright East Malling Res. Stat.

20. Growth and form influenced by technique

One-year trees of Conference pear on Quince A grown side by side. Two bundles (left) budded in July; two bundles (right) whip-and-tongue grafted the following March. All material of equal source, worked by one operator. Lateral scion-shoots have been removed from grafted trees.

grafts have been used, but the popular methods in use to-day are few. Once the operator is well aware of the position of the cambium and realizes the vital necessity of cambial contact (Chapter II), the actual grafting presents no problem. The preparation of the trees to be grafted and the position and aftercare of the grafts is less well understood and is often the cause of mediocre results or even of complete failure.

SEASON FOR GRAFTING

Spring is the normal time for reworking trees with dormant scions, though grafting in various forms is done almost throughout the year. The methods to be described here are intended for use in England from February to June. Early grafting usually gives the best results, though some forms of grafting with dormant scions are successful up to mid-summer and after.

The rind of stone fruits, such as plums and cherries, does not readily separate from the wood until very late in the season, hence they are usually cleft grafted.

PREPARATORY MEASURES

Ample supplies of scion-wood must be collected and stored, and graft seals and tools made ready before grafting time, as described in Chapters IV and V.

Steps and ladders should be provided as required. Ladders often prove less suitable than steps because of the danger of damaging the new scions by sliding along the boughs. Four-wheel trolleys make excellent platforms and these may be augmented at will by placing on them boxes, tubs or steps. Such trolley platforms are conveniently mobile and are readily moved around and from tree to tree, besides transporting the paraphernalia of the workers.

Each grafter should be equipped with a satchel, which should be worn well up on the chest. The knife should be placed in a separate part of the satchel where it is clearly visible, so that it can be left open with safety. There should be a pouch for scions well up, almost under the chin. These arrangements, which may be combined in an apron (Fig. 105), help to expedite the work of the grafter.

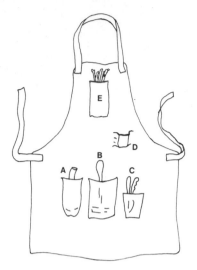

Fig. 105 Grafter's apron

Of strong canvas or leather, reaching barely halfway down the thighs; useful when topworking and frameworking. A wad of cotton wool is placed in the bottom of the knife pouch. Knife (A), hammer (B), and secateurs (C), firmly gripped in their pouches, protrude for ready handgrasp. Gusseted finger-and-thumb pouch for small nails (D). Scion pouch well up on the chest (E).

Topworking

GENERAL

Much of the main branch system of the tree is removed and scions are inserted in the cut ends of the main limbs, within 600–900 mm (2–3 ft.) of the crutch of medium-sized trees. A well-prepared tree is shown in Plate 21. Correspondingly more framework should be retained in larger trees. All small branches below the general level of grafting should be left intact for at least a year in order to feed the roots and retain life in the tree. Such branches are commonly known as sap-drawers. Experience has shown that, where there are insufficient sap-drawers, one or two large branches of the old variety (Plate 22) should be left intact for the same purpose. These may be grafted one or two years later, or removed entirely when growth from the new scions is abundant. The presence of large sap-drawers in no way diminishes scion growth but, on the contrary, increases it, encourages healing,

21. A twenty-year-old apple tree suitably prepared and topworked

Note the presence of numerous small sap-drawer branches and the lead-arsenate protective dressing on the scions to prevent weevil damage.

22. Sap-drawer branches

Two complete branches retained one or two years in a topworked tree. Such branches keep the tree healthy, aid the healing of the wounds and the formation of strong unions. They should be removed when scion growth is abundant.

and reduces the risk of loss due to such diseases as silver-leaf (*Stereum purpureum*) which enter through wounds[41,131].

The main limbs of the trees to be grafted may be sawn off at any time during winter, but fresh-cut surfaces at grafting time are considered an advantage. The preparation of the tree involves considerable labour, and some growers prefer to do this in the winter so that the lop-and-top may be cleared away when labour can best be spared. On the other hand, large wounds should not be made whilst trees are completely dormant, especially in stone fruits and certain apples such as Newton Wonder and Early Victoria, susceptible to silver-leaf. Very severe pruning in winter also has the effect of delaying the start of growth in the spring.

In preparing trees for topworking it should be remembered that large wounds take a long time to heal, and it is better to insert scions into two small limbs above a fork rather than in one large limb lower down. Branches of 125 mm (5 in.) or more in diameter should be sawn off higher up, where their diameter is not more than 75–100 mm (3–4 in.). Eventually it is best if only one scion remains at the end of each limb of the old variety, as this makes for strength at the union. Two or three scions may be inserted in order to retain life at the edges of the cut surface, but all except one should be removed as the wound heals.

The methods of grafting used in topworking are very numerous. Only the chief of these are described here, namely, two methods of cleft grafting—simple and oblique—and three rind methods.

CLEFT GRAFT (*split graft*)

The latter part of February is not too early to begin cleft grafting and the work may continue as long as the scion-wood remains in good condition, having in mind that in general early grafting is preferable. The cut end of the limb is cleft diametrically. This cleft is opened by the insertion of a wedge or by using the grafting tool as a lever. The scions, usually of three or four buds, are prepared by two slanting cuts at their base in the form of a long, tapering wedge (Fig. 106, A). The presence of a bud (B) between the two cut surfaces and about half-way between the point at which the cuts begin and the

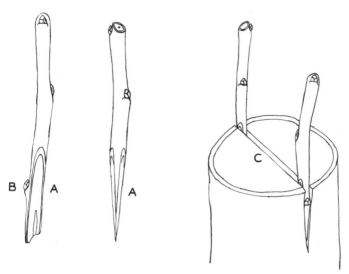

Fig. 106 Cleft graft (split graft)

base of the scion is an advantage. The scions are inserted so that their cambial regions are in contact, due allowance being made for the thick rind of the stock and the thin rind of the scion. The wedge or tool is removed and the scions are held fast by the pressure of the limb of the stock (C). Should only one scion be inserted in the end of each limb, it is unnecessary to extend the cleft right across the limb. It should then extend only halfway across and should be sufficiently deep to allow for opening and the insertion of the scion.

For rapid work it is an advantage to sling the tool and mallet together (Fig. 107) with a thong or cord, so that when the tool is used as a wedge the mallet may be left to dangle and the operator is free to use both hands for the preparation and insertion of the scions. The work proceeds as follows: the tool is placed on the end of the limb, one powerful blow drives it well home, the tool is wrenched out and the wedge-like end is driven down in the middle of the cleft, the mallet is dropped and dangles. The scions are then prepared and inserted, and with one upward blow the tool is removed.

When sealing large clefts it appears to be an advantage to place a little clay both in the clefts, to prevent liquid seals

253

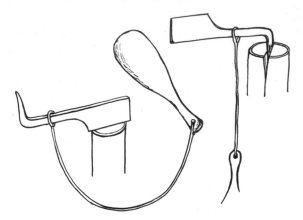

Fig. 107 Tool and mallet slung together

The cord, or thong, should be long enough to permit free movement. When the wedge is driven into the cleft, the mallet is dropped and the operator uses both hands to prepare and insert the scion.

from running too far into them and between any ill-fitting scion and stock.

Though often successful, this somewhat brutal technique is now being replaced by the oblique cleft method, in which the heart of the limb is not split.

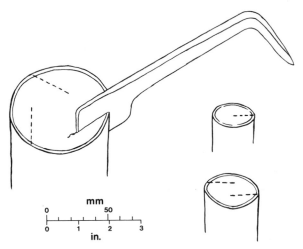

Fig. 108 Methods of cleaving obliquely

Note treatment of large, medium, and small limbs.

OBLIQUE CLEFT GRAFT

The sawn ends of the limbs are pared smooth, at least near the edges, and cleft as shown in Fig. 108. None of the clefts should extend right across the branch.

Three and four-bud scions are used. The upper end is cut immediately above a bud and the lower is prepared by two slanting cuts in the form of a long tapering wedge as for the ordinary cleft graft (Fig. 106, A). Again the presence of a bud (B) between the two cut surfaces and about halfway between the point at which the cuts begin and the base of the scion is an advantage.

The clefts are opened by means of the grafting tool and the scions are inserted, care being taken to see that the cambia are in contact (Fig. 109). The exposed cambium of the scion is not strictly in line with the scion, so a slight inclination outwards of the top of the scion gives more opportunities for the cambia to meet. Removal of the tool causes the scion to be held firmly in position and no tying or nailing is necessary. All exposed cut surfaces, including the apical ends of the scions, and the cut surfaces of limbs, are sealed with grafting-wax. Careful workers rub a little clay into the clefts to prevent the wax from running down inside, where it might prevent healing.

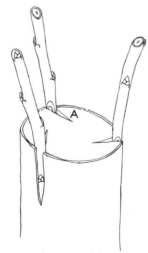

Fig. 109 Oblique cleft graft

RIND GRAFT (*crown graft*)

An example of crown grafting is shown in Plate 23.

The sawn ends of the limbs are pared smooth and the scion is inserted between the rind and the wood of the trees. The rind should part readily from the wood, a condition not

23. A standard apple changed to a bush by crown grafting

Fourteen veneer crown grafts (nailed) form the crown. A dozen inverted L grafts are used to provide additional early leaf activity. The tree should be sprayed with whitewash from the sun side to reflect midday heat. On the right, a four-year-old decapitated specimen with circumference healed and at the branch-thinning stage.

normally reached in England until after early April.

There are many methods of preparing and inserting the scion. In the most common, the scion is prepared as for whip-and-tongue grafting but without the tongue (Fig. 57). The prepared limb receives a cut of one or two inches in length extending downwards from the pared surface of the limb, and the scion is pushed down beneath the rind, so that the cut surface of the scion is towards the middle of the limb (Fig. 110). It is sometimes better, particularly when the rind of the stock is very thick, to lift the rind only on one side of the cut and to remove a very narrow shaving from the edge of the

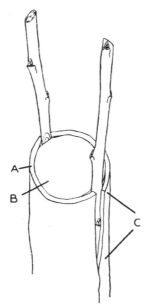

Fig. 110 Rind graft (crown graft)

scion (Fig. 111), which is against the uplifted rind. The rind may be first lifted by means of a thin bone spatula, such as can be made from an old toothbrush handle. The limb receives from one to four scions according to its size.

Unlike the oblique cleft the rind graft must be tied firmly with soft string, raffia, or tape around the end of the limb over the inserted sections, avoiding the stock or basal bud of the

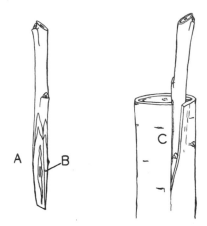

Fig. 111 Modified rind graft

The scion is prepared by one principal cut (A) and one shallow straightening cut (B). The rind of the stock is raised on one side only of the incision and the shallow cut (B) is placed against the edge of the non-raised rind (C).

scion. If waxed tape is used for binding, it will only be necessary to seal the remaining exposed cut surface at the end of the branch and the tips of the scions: otherwise all cut surfaces, including the bark opening over the inserted scion, must be sealed as in oblique cleft grafting.

Soft string is usually more convenient to use than raffia, especially on large limbs. Raffia and some adhesive tapes usually burst with the growth of the tree, but to avoid constriction the string may have to be cut by the middle of the summer. The work should be inspected from time to time.

VENEER CROWN GRAFT (*Tittel's graft*)

In the veneer type of rind graft the scion is prepared as for normal rind grafting with additional small cuts (Fig. 112, A) at the sides, just sufficient to straighten the edges. The thin tip at the base of the scion is also removed. The transversely cut limb of the stock receives two parallel cuts separated by the width of the prepared scion. The scion is pushed down beneath the rind lying between the parallel cuts. The strip of the stock rind (B) is pushed out in this process and is cut off flush with the limb. The graft may be tied or fixed by two

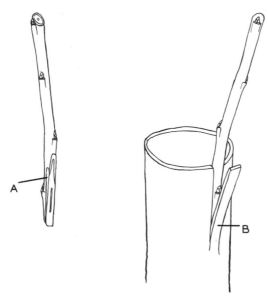

Fig. 112 Veneer crown graft (Tittel's graft)

16 mm ($\frac{5}{8}$ in.) gimp pins driven through the scion into the stock. Sealing is necessary.

When the stock rind is very thick the scion, when inserted, may not rise above the surface and tying will not hold the scion. In such a case the stock rind should be thinned by paring away the outer corky layers before inserting the scion[111]. Such treatment is not necessary where nails are used for fixing.

STRAP GRAFT
This is an excellent method for the topworking of trees, largely because it aids the healing of the ends of the limbs and also serves to 'tie' the scion to the stock. The comparative tediousness of the method must be set against the foregoing advantages.

The scions are prepared by first raising a strap of rind (Fig. 113, A) long enough to stretch across the cut surface of the stock and an inch or so down one side. This strap must be thick enough to contain a continuous layer of cambium but

259

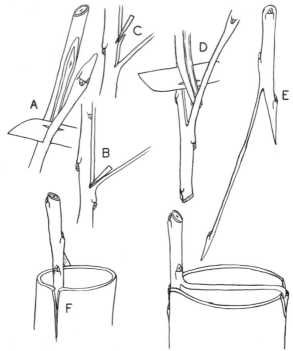

Fig. 113 Strap graft
Careful tying and sealing are necessary to complete this graft.

not so thick as to be stiff. It must not snap when folded sharply to form a right angle. When the strap is long enough the knife is turned slightly deeper into the scion (B), and a few shavings are taken from the scion just below the upper end of the strap (C). The knife is reversed to slice the scion after the manner of a splice graft scion (D). The ideal scion is cut so that a 'stock' bud (E) is left opposite this slicing cut. The end of the limb to be grafted is pared smooth and the rind is slit (F) as described for rind grafting. The prepared scion is pushed down under the stock rind, taking care that the strap is not torn away as the scion is pushed home. The strap now lies flat upon the cut surface at the end of the stock, and the free end is either tucked under the rind or a slice of stock rind is removed to expose the cambium upon which the end of the strap is pressed, the whole graft being tied with soft string or adhesive tape, sealing being done in the usual manner.

Large limbs may receive two scions, the straps lying parallel, and occasionally three, or even four, straps may be used so that they cross one another. Small limbs, young trees, and root-stocks in nurseries may be strap grafted after the manner of the Somerset saddle graft (page 185).

Successful strap grafts heal rapidly and form immensely strong junctions.

AFTERCARE OF TOPWORKED TREES

Scions newly inserted when topworking are often damaged by weevils which eat the rind and the buds. Such pests may be controlled by painting the scions with a cream of lead arsenate paste and water immediately after grafting.

If the trees are in an exposed position the 'leaders', or main shoots from the scions, may be steadied during the first year by tying them loosely to stakes bound lightly to the branches of the tree.

Grafts which are tied with twine or other non-perishable ligatures must be watched and released before constriction occurs by passing a sharp knife through the strands between the scions. Unwrapping should not be necessary and may unsettle the scions.

Topworking is followed by profuse and vigorous sucker growths from the stock. These growths may be thinned but not removed, except where they tend to smother the scions completely. If the scions grow very vigorously in their first season, the stock suckers may be removed the following winter or very much reduced in quantity; otherwise some should be left for a further season. Similarly the removal of the sap-drawers (page 250) should be governed by the growth from the scions.

At no time should the limbs of the tree (the stock) be left unshaded, as such exposure quite often results in injury known as sunscald. This damage occurs on the surfaces of limbs facing the noonday sun and can be very severe, even in temperate climates. Where the grafted limbs are bare of branches, and shade from sap-drawers is inadequate, the surfaces of the limbs facing south should be whitened. Where a number of trees require this protection it may be most easily given by spraying with thick limewash or whitening from a southerly direction. Such treatment may check temperature increases in sunny weather. Interesting observations[13,17] have been made concerning the prevention of sunscald. Trees with thick, corky bark are not susceptible, and until this cork develops, some trees, notably certain conifers (pines), have light-coloured bark. Scald is common in trees suddenly isolated. Rapid growth from the scions may help to prevent excessive heating by developing an efficient transpiratory stream which may act as a cooling system beneath the rind. Where trees are affected with sunscald, the limbs are red-brown when viewed from the south and normal green or grey from the north.

The actual operation of topworking is only the beginning of the conversion of the tree. For some years the sucker growths must receive attention and the scion growths must not be neglected, but pruned to form a sturdy branch system, much as advocated for vigorously growing young trees. Where more than one scion is growing from each limb the better placed must be encouraged by reducing the others until they are eventually entirely removed. Unhealed wounds should be resealed to reduce the danger of the heart-wood rotting (Plate 24). When scions have failed to grow, the stock should be allowed to put forth shoots close to the end of the limb to prevent dying back from the cut end. These limbs may be regrafted, or cut off close to the trunk, the following spring. Topworked trees cannot be expected to carry crops until they have developed a substantial framework.

Frameworking

GENERAL

The initial cost of frameworking is much higher than top-working, but there are some important advantages. Frame-worked trees quickly yield large crops of good-quality fruit (Plate 25). Pears, plums, and cherries have given useful crops in the third season from grafting. Frameworked trees remain healthy in situations where topworked trees may die from diseases such as silver-leaf and papery bark canker[52].

In preparing trees for frameworking a few badly placed branches may be removed entirely, but this work should be

24. An unhealed and weak crotch

One of the scions on the end of this limb should have been reduced in favour of the other so that, finally, there would be no weakness at the bifurcation. The large wound should have been resealed until healed.

25. Frameworked trees crop quickly

A 'Newton Wonder' apple tree, frameworked at thirteen years old with 'Laxton's Superb', bearing five bushels (about 90 kg) of good-quality fruit two and a half years from grafting.

kept to a minimum in order to avoid making large wounds and disturbing the natural growth balance of roots and shoots. The small lateral branches and spurs must be pruned away, except in stub grafting, and are replaced at every 200–250 mm (8–10 in.) along the limbs, right out to the two-year wood, with scions. In this way practically the whole of the existing framework of the tree is retained and only the fruiting laterals and young extension shoots are replaced with the new variety (Plate 26).

Copyright East Malling Res. Stat.

26. A frameworked tree

A twenty-year-old 'Gladstone' apple tree immediately after frameworking with 120 scions of 'Worcester Pearmain'. Note that the whole framework of the tree is retained and only the lateral branches are replaced with the new variety.

LENGTH OF SCIONS

In all methods of frameworking long scions must be used, never having less than six buds and preferably seven or eight. Growth from newly inserted scions is almost always extremely vigorous and scions containing four or fewer buds, as used in topworking, produce strong shoots from each bud which crowd the tree with unfruitful wood. Long scions, containing seven or eight buds, grow less vigorously and form

27. Long scions for frameworking

Short scions (on left) give rise to vigorous shoots which crowd the tree with unfruitful wood. Scions with seven to nine buds (on right) form shoots from the upper two or three buds and produce fruit spurs or short fruiting laterals from the lower.

shoots from the upper two or three buds only (Plate 27), and produce fruit-buds or short fruiting laterals, from the lower buds.

POSITION OF SCIONS

In order to obtain an even growth from all the scions great precision is required in the placing of grafts. Scions in a vertical position, especially on the inner and top surface of the limb, make stronger growth than those approaching a horizontal position. Very vigorous growth is not needed except where an extension branch is required. Scions on the under side of the limb and in a horizontal position produce the weakest growth. An even scion distribution, placed so as to obtain even vigour comparable with that of a normal tree, is to be preferred.

The placing of scions near the crutch of the tree does not permit of their proper development and they should, therefore, be at least 450–600 mm (18–24 in.) from the crutch. Higher up the limbs where it is necessary to put a few scions on the inside of the tree, they should be placed in a lateral position so as to avoid the development of strong growth. Again, scions should not be placed in low and outer positions where they may be damaged during cultural operations.

Where stub grafting is used entirely, there is sometimes a risk of having stubs on one side of the limbs only. This gives an unnatural and one-sided development. Stub grafts should only be used where the position is suitable, and the gaps should be filled with side grafts or inverted L rind grafts, according to the season.

STUB GRAFT

The tree to be grafted is cleared of all lateral growths, except small lateral branches 6–25 mm ($\frac{1}{4}$–1 in.) in diameter, which are retained wherever it is desired to have lateral branches of the new variety. This work may be carried out at any time during the dormant season prior to grafting. Where there is definite danger of infection by the silver-leaf fungus (*Stereum purpureum*), it is considered somewhat safer to defer preparing the trees until the time of grafting.

The scions are prepared with their basal ends cut in the form of a wedge (Fig. 114, A). One side of the wedge receives a slightly longer cut than the other. The lateral branch to be grafted receives a cut on its upper side beginning about 12 mm ($\frac{1}{2}$ in.) from the main branch and extending to the base, almost to the centre of the lateral shoot. Care must be taken to ensure that the cleft does not extend beyond the centre of the shoot, otherwise the 'spring' of the cleft is weakened and the scion will not be held sufficiently firmly. This cut is opened by bending down the lateral, and the prepared scion is inserted with the longer side of the wedge downwards (B). It is essential that at least a portion of the cambial tissue of the scion should be in contact with that of the lateral branch into which it is inserted. In order to achieve

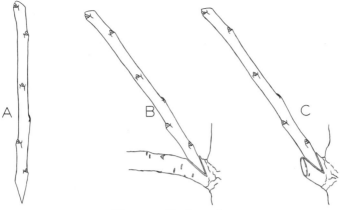

Fig. 114 Stub graft

this when inserting scions into large laterals, it is necessary to place the scion to one side of the cleft. The lateral is released and cut off immediately above the insertion of the scion (C). Upon first attempting this operation, some difficulty may be experienced, but with a good knife and after a little practice it can be satisfactorily accomplished. Some use secateurs, but a knife is preferable. The terminal shoot on each main branch is cut off at the time of grafting at a point immediately above the uppermost stub graft. The sealing of all cut surfaces completes the operation.

SIDE GRAFT

Here scions are inserted direct into the limbs of the tree without recourse to the use of lateral shoots. Comparatively small limbs are more easily side grafted than those of a large size. The trees are prepared by removing all the lateral shoots before grafting and the whole framework is then worked with side grafts. Alternatively, the best-placed laterals may be left and stub grafted, and in that case only the intermediate portions of the limbs are furnished with side grafts.

The scions are prepared with the basal ends in the form of an unequally-sided and comparatively long wedge. In preparing the scions advantage should be taken of the slightly zigzag nature of the shoots, in order that the scion, when

267

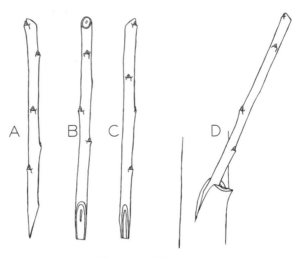

Fig. 115 Side graft

inserted, may stand away from the framework branch as much as possible, so that the angle is approximately that of a lateral shoot. The prepared scion is depicted in Fig. 115, A, B, and C. A cut is made into the side of the branch at an angle of about 20° and never deeper than one-quarter of the diameter of the branch. The cut is opened by slight bending of the branch and the scion is inserted so that its cambium is in contact with that of the framework. The thin lip of bark raised from the branch prior to inserting the graft is cut off after the scion has been pushed home (D). In order to complete the graft it only remains to seal the exposed cut surfaces.

OBLIQUE SIDE GRAFT

The scion is prepared (Fig. 116, A and B) with two sloping cuts about 25 mm (1 in.) long made on two sides of the basal end of the scion, and in such a manner that the two cuts meet longitudinally on one side at an angle of 30° to 45°. A shallow, oblique cleft is made in the side of the limb (Fig. 116, C). This cleft should be slightly deeper than the lateral wedge at the base of the scion. The knife is then partly removed from the cleft to allow the toe of the scion to be inserted. When the scion has entered the cleft, the knife is removed and the scion

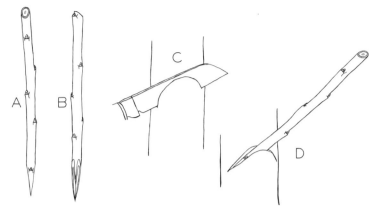

Fig. 116 Oblique side graft

wedge is then pushed right across the cleft. The operation is completed by sealing the wound. This side graft has proved to be very suitable for working large limbs where the rind is often very thick.

It is particularly important to ensure that in side cleft grafting, as in other methods, the cambial tissues of scion and stock are in contact with each other.

INVERTED L RIND GRAFT

Rind grafting must be done when the rind is lifting. All lateral shoots and spurs are removed from the tree before grafting begins. The scion is prepared by first making one sloping cut of about 40 mm ($1\frac{1}{2}$ in.) in length at the basal end, followed by a very shallow cut at one side, just sufficient to expose the cambium (Fig. 117, A). The scion is then reversed and a shallower cut than the first is made a little to the side of the scion, away from the small second cut (B). The limb to be grafted now receives a cut as depicted in Fig. 117, C. The term 'inverted L' does not accurately describe this cut. In the diagram it will be noted that the longer, or lower, cut is not made in line with the limb, as this would cause the scion to lie along the limb at an unnatural angle, but it is so made that the scion stands away from the surface of the limb. The top cut in the rind is not made at right angles to the lower one but more nearly approaches 150°. It is also cut obliquely into the rind,

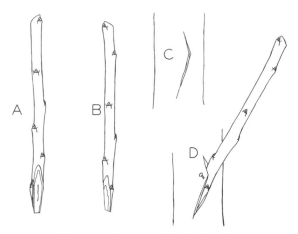

Fig. 117 Inverted L rind graft

thus enabling the scion to enter beneath the rind more easily and to fit snugly into position. This modification of the inverted L method allows the rind to fit closely to the scion. The scion is inserted (D) and is held firmly in position by a flatheaded nail or gimp pin, No. 20 gauge and 16 mm ($\frac{5}{8}$ in.) in length, which is driven right home through the rind and scion into the wood. When the rind of the limb is abnormally thick, 20 mm ($\frac{3}{4}$ in.) nails of the same gauge are used. It has been found that stouter nails tend to split the scion. Scions may be placed immediately below wounds caused by the removal of lateral shoots or spurs (E). Scions so placed assist the rapid healing of the wound and the sealing of the graft seals the wound, thus saving both time and material. In sealing this type of graft it is especially important to ensure that water cannot enter between the scion and the limb.

SLIT GRAFT

This is a simplification of the awl or needle graft used in various parts of the world. It is successful on apples and pears, where the rind separates readily from the wood. Branches less than 25 mm (1 in.) thick are not suitable for slit grafting and should be stub or side grafted.

The tree is cleared of laterals. The base of the scion is cut to

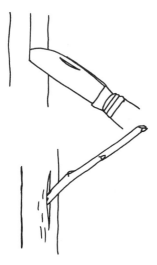

Fig. 118 Slit graft

form a wedge about 25 mm (1 in.) long. A 50 mm (2 in.) oblique slit is made along the limb so that the tip of the knife meets the wood of the limb at a tangent (Fig. 118).

The scion is pushed into the slit and finds its way between wood and rind. The scion may be set at any desired angle by raising or lowering the apex.

All cut surfaces are sealed; no tying or nailing is required.

AFTERCARE OF FRAMEWORKED TREES

Should the clay-coloured and closely-related weevils prove troublesome by eating the rind and buds on the scions, the tree should be grease-banded in the normal way. This prevents those species of weevil from crawling up the trunk from their haunts on the ground. The green and brown leaf-eating weevils, which attack developing leaves, all fly readily; one or more applications of derris will control them.

In clean-cultivated orchards, where frameworking has been completed, the soil will be in urgent need of cultivation owing to the treading of the grafters. Care should be taken, during the first few cultivations, that the trees are not shaken severely by horses or tractors, as this may dislodge some of the cleft grafts before they become firmly established.

As the season advances, sucker shoots will arise from the main branches of the old variety, and all these should be rubbed out when the longest are less than 100 mm (4 in.) in length. By this time most of the sucker shoots likely to arise can be removed and it may not be necessary to remove any more until the annual winter pruning, but, if necessary, the operation should be repeated during the summer. If sufficient scions have been used, very few sucker shoots will arise in later years, and these should be removed at pruning time.

Some growers have found the pruning of frameworked trees somewhat of a problem. This pruning may begin immediately the grafters have left the tree. Scions suitably placed for leaders may be shortened to half their length, but this early treatment is not recommended except to the most careful workers who are able to guarantee not to move the scions and to seal the pruning cut as in grafting. More usually the first pruning takes place in the winter following grafting and can be considered in three steps:

1. Any sucker shoots and crowding scion growths should be removed. There should be no hesitation in sawing away whole scions where these are badly placed.

2. Shoots selected as leaders, for extending the main branch system or to form new branches, should be tipped by removal of a quarter or a third of their length.

3. Of all the remaining shoots about a third should be shortened to half their length.

The frameworked tree lends itself to the 'renewal' system of pruning, details of which may be found in Bulletin No. 133, *Apples and Pears* (now No. 207, *Apples* and No. 208, *Pears*), published by H.M. Stationery Office.

PROVISION OF POLLINATORS

Frameworking has proved to be a rapid and highly successful means of regrafting whole trees of apples, pears, plums, and cherries. Further, particularly in the stub graft, it provides a useful method of supplying single pollinator branches in individual trees of an established variety[38].

It is quite common to find orchards or plantations of single varieties. Bramley's seedling apple has frequently been planted without providing for cross-pollination, and this variety cannot be relied upon to set regular crops unless other varieties, which flower at the same time, are near enough for bees and other insects to carry pollen easily to the Bramleys even in cold weather.

Experience teaches the value of cross-pollination, and in cases of poor cropping, where other conditions are favourable, the introduction of other pollen would almost certainly greatly improve returns. Pollinator varieties should be chosen which blossom freely and annually, and at the same time as the main variety. It is important that they should withstand sprays required by the main variety.

In the absence of experimental evidence the optimum proportion of the pollinator is a matter for speculation. It has been suggested that one whole tree in nine is adequate, but the distance of planting is of some importance and it is wise to reckon distance between pollinators as well as quantity. A branch of pollinator at every 9 m (30 ft.) is possibly a reasonable allowance. In a 4·5 m (15 ft.) plant, this would mean a part of every other tree or a branch on every tree at 9 m (30 ft.) square. With ten scions per branch this is about 1200 scions per hectare, or rather less than 500 scions per acre. A grafter and waxer can complete four branches per hour, thus requiring 30 hours per hectare, or 12·5 hours per acre. Using two ladders the grafter leaves his first ladder in position for use by the waxer, and proceeds to the second ladder, previously placed in position by the waxer.

TO REPAIR AND IMPROVE

The use of grafting in so-called tree surgery could well be extended for the strengthening and general improvement of many trees, both ornamental and economic. If we observe the rules of cambial contact and polarity and, of course, relationship, there seems to be little limit to the use of this repair grafting.

Bridge Grafting

When trees have been completely girdled by animals, implements, or canker, they die unless branches and roots are reconnected. Death may not occur until one or two years after

273

28. Girdled tree saved by bridge grafting

*This thirty-year-old apple tree, when 250 mm (10 in.) in diameter, was
completely girdled by rabbits and would have died had it not been bridge
grafted. The tree has continued to grow and crop normally. Photographed
(left) six years and (right) thirty years after bridging.*

girdling, since nutrients pass up in the undamaged wood and
enable the leaves to function, but eventually the roots die
because no food returns to them via the rind. Reconnection of
the girdled rind saves the tree (Plates 28 and 29). Girdling is
not always complete, either around the tree or down to the
wood, and in these cases bridge grafting is not necessary, but
when in doubt it should be done and regarded as an inexpensive form of insurance. If the rind has been eaten away
down to ground level, the soil should be removed to expose
sufficient undamaged surface and replaced after grafting.

The rough edges of the bark may be trimmed, and if
girdling is due to canker, all discoloured tissue must be pared
away before grafting.

Dormant scions of sufficient length for ease of manipula-

29. Hawser-girdled oak (*Quercus robur*) saved by bridging
A. One year after damage. Note very thick bark. B. Bark channelled to reach flexible inner rind. C. Young seedling oak scions veneer grafted, nailed and sealed. D. Six-year-old bridge. Girdle remains unhealed. E. Position of failed bridge.

tion are required. If they are not available, freshly-collected defoliated shoots of the past season's growth may be used. When the girdled area is of great length the scions may not be long enough and it will be necessary to splice two scions together by whip-and-tongue grafting. Small trees about 25 mm (1 in.) in diameter can be saved by one bridge. Larger trees should have one bridge for every 25 mm (1 in.) of diameter. Thus a trunk 150 mm (6 in.) thick would have five or six bridges.

There are many ways of inserting the scions. One of the simplest which has proved successful is the inverted L rind graft (Fig. 117) used in frameworking. Insert the lower end first a short way below the girdle and nail firmly in position; then prepare and insert the apical end the same distance above the girdle in a reversed incision. Short scions and those that are extra stout are difficult to manipulate. It is easier to obtain a good fit if the bridge is bowed outwards, and this is aided by inserting a tapered piece of wood, 40–50 mm (1½–2 in.) thick, between the tree and the bridge whilst the graft is being nailed home. The necessity for bending the bridge outwards is not always appreciated by beginners. Advantage should be taken of the zigzag nature of scion-wood and the principal cuts on the scion made by slicing away a bud at each end of the bridge (Fig. 119, A). One end being fixed, bending (B) of the scion back to the stock will bring the cut surface of the scion parallel with the cambial surface of the stock (C). Preparing the scion in any other way will not achieve this end.

Fig. 119 Bridging by rind grafting

As in the side graft (Fig. 115) advantage is taken of the zigzag nature of the scion-wood. The scion is first fixed at its base (B) and then the upper end is bent to the stock (C) and similarly secured (see text).

Very small trees are not conveniently rind grafted. These may be bridged by side cleft grafting as advocated for side grafting small branches in frameworking (Fig. 115). Here the cleft is not made so deep into the wood and the graft requires tying firmly with tape, string, or strong raffia.

All cut surfaces should be sealed, taking care that no holes are left for the entry of water between the bridge and the tree at the graft.

Immediately each tree is grafted it should be protected from further damage. Efficient wiring of the whole piece will prevent damage by rabbits, but tree-guards will be necessary to protect orchard trees against grazing animals. Implements and spray hoses cause damage to bridges, particularly whilst young, and stakes should be used to guard against this.

Young growths from the bridges should be rubbed out as

Courtesy of E. B. Mannington, Esq., Kent, England

30. Cherry trees attacked by bacterial canker saved by bridge grafting

A diseased crotch induces stem suckers which are grafted into the healthy branches. Left : six weeks after grafting. Right : another tree (bridged five years earlier) still cropping.

they appear, and the tying material used on small trees adjusted or removed before constriction occurs.

Cherry trees partly or completely girdled by bacterial canker (*Pseudomonas mors-prunorum*) have been saved by bridge grafting during the growing season (Plate 30). Bacterial cankers almost always complete their development in one season and any extension of damage is generally due to reinfection, so there is little advantage in trimming dead or dying parts. A diseased stem is normally first noticed following the growth of young, vigorous shoots from below the canker, due to girdling. These shoots grow rapidly and make excellent bridges if inarched above the canker. If sprouts are not available, well-developed, firm shoots are collected from healthy rootstocks, or vigorous young trees, for use as bridges. These should be defoliated but the leaves on inarched sprouts should be left, for they will not dry out the bridge but help to sustain the lower parts of the tree. As the work is best done in late spring or summer it is convenient to use rind methods of grafting.

BRIDGING INCOMPATIBLE UNIONS

Some forms of incompatibility between stock and scion, such as that exhibited by a clean break at the union, may be overcome by bridging the union with a piece of another variety known to be compatible with both. This treatment has proved efficacious in trees of Williams' Bon Chrétien pear worked direct on quince[39]. Sometimes these trees grow and crop normally for many years, but usually there is considerable variation in the trees of a single planting. This may extend from the normal tree down to the tree that breaks clean off at the graft union a few years after planting. These losses are prevented by double-working in the nursery (page 233) and bridging should be regarded merely as making the best of a bad job.

The first essential is to *stake thoroughly* so that the strain on the union is reduced. There are many methods of staking; slanting stakes are good but should be attached to the tree at a high level to lessen the purchase of the top. Vertical stakes, especially if attached at two places, are probably ideal providing rubbing can be avoided.

Almost all trees planted in recent years have been set with their unions above ground level in order to prevent scion-rooting[62]. In these cases there is room to bridge the union by inserting an intermediate into the rootstock, but if the union is not clear of the ground the soil must be pulled away and not replaced by future cultivations, otherwise the intermediate bridge may root, usually putting the tree into a non-fruitful condition. Where the unions are deeply buried, bridging is not a practical proposition and in many cases the tree will have already scion-rooted.

To bridge trees in which the union is close to the ground a composite bridge of rootstock and intermediate, spliced together, is used (Fig. 120, A). A novel method of increasing the strength of incompatible unions of the nature described is to insert panels of rind of the chosen intermediate across the line of the union (B), taking care to insert such panels the right way up. Three or four panels about 12 mm ($\frac{1}{2}$ in.) wide should

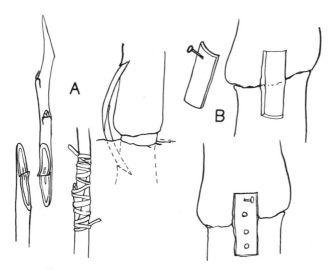

Fig. 120 Two methods of bridging unions

A. When the union is close to the ground the risk of scion rooting from the bridge is overcome by the use of a spliced bridge, the lower part being the same variety as the rootstock upon which the tree is worked. B. An incompatible union may be neatly 'bridged' by veneering panels of rind, of a suitably compatible variety, across the single-worked junction. These panels grow sectorially and 'staple' the original scion to the rootstock.

prove sufficient for trees up to 75 mm (3 in.) in diameter. These lay down woody strands which, uniting with both scion and stock, staple the existing junction.

Beurré Hardy and Fertility have been used for bridging single-worked trees of Williams' Bon Chrétien on Quince A. The scion-wood is collected in the winter, and in March or April is inserted into the quince rootstock immediately below the union and into the scion variety a short distance above. The bridge should be as short as can be manipulated; this is usually about 200 mm (8 in.). Small trees two or three years old might receive one bridge, but two bridges give more chance of success with larger trees. There are various ways of inserting the bridge, but a simple side cleft appears to be all that is required (Fig. 121). A shoot of the intermediate variety

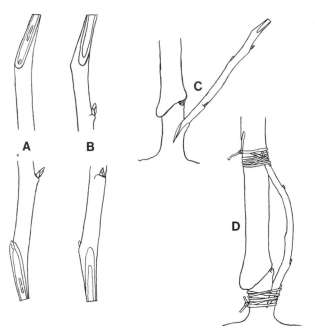

Fig. 121 Bridging incompatible unions in small trees

A. The bridge showing cut surfaces to go towards the tree. B. Shallow cuts on the outside of the bridge to expose the cambium. C. The bridge inserted into the quince rootstock below the union. D. The bridge firmly tied and ready for sealing with grafting wax.

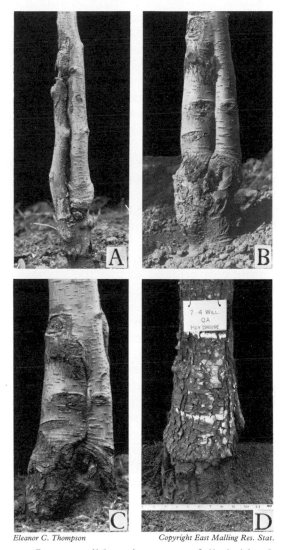

Eleanor C. Thompson   Copyright East Malling Res. Stat.

31. Incompatible union successfully bridged

*A. Two years after bridging (bridge on the left). B. Five years after bridging.
C. Four years later. The bridged portion of the tree may now be sawn away to
complete the conversion from a single to a double-worked tree. D. A twenty-
four-year-old bridge has completely replaced the original trunk, which it has
thrust aside.*

is cut into a comparatively long wedge by slicing away a bud and removing sufficient bark on the opposite side to expose the cambium. A cut is made downwards, immediately below the union, into which the base of the scion is inserted. This is now tied very firmly, with filis (soft twine) or tape, after which the upper end of the bridge is prepared and inserted similarly, but in the reverse direction, into the trunk above the union. Firm tying and thorough sealing of all cut surfaces is important, taking care that no holes are left for the entry of water between the bridge and the tree at the graft. Bridging the unions of large trees has not been tried, but it seems likely that the rind graft, as used in frameworking, might be used in such cases with success.

Young growths from the bridges should be rubbed out as they appear and the tying material adjusted or removed before constriction occurs.

Successful bridges rapidly increase in size and soon begin to overtake the part bridged so that the single-worked tree becomes double-worked (Plate 31).

Inarching and Approach Grafting

Where the use of unsuitable rootstocks or damage to the roots has resulted in stunted trees, it is often possible to improve growth by inarching into them more vigorous rootstocks which eventually take over the duties of the original roots[63,66].

Two rootstocks placed on opposite sides of the tree should be sufficient (Plate 32). From recent experiments it appears that the stocks should be inserted in line with the crutch between the major branches in order to hasten the distribution of the effect of each of the inarches to more than one branch. Trunks of trees frequently exhibit torsion and the lines of growth should be followed from the crutches downwards in choosing such points for inarching.

The inarch should be made as low as convenient. The rootstock may be planted either during the winter or at grafting time in April. In order to hasten the growth of the new roots a fairly large hole should be opened, between the existing main roots, and filled with top soil in which has been incorporated some rotted dung or humus. The rootstock should be planted firmly. If planted at grafting time the graft

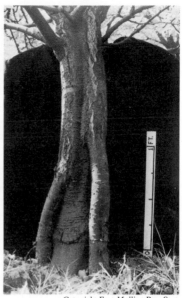

32. Inarching a plum tree

Top left : A tree broken off clean at the graft junction and showing the original tongue made in the stock when grafting. Below left : fellow trees were inarched with small rootstocks, known to be compatible with the scion variety, and these rapidly increased in size and importance to the tree. Right : photograph taken five years after inarching.

may be completed before placing soil around the rootstock. The method of insertion is as described for the apical ends of bridge grafts, and inarches should also be similarly protected and disbudded. Various methods of inserting inarches are shown (Figs. 31, 32, and 33).

Mature trees often have extremely thick rind and consequently manipulation by rind methods is difficult. Hodgson[66], who describes the inarching of citrus, advocates slicing off the outer layers of the rind of the trunk at the place of inarching. This should be done before incising the rind. He writes: 'The thinning of the bark accomplishes several objects; it renders it pliable so that separation from the cambium does not occur except in the immediate vicinity of the incisions, a matter of importance; and it cleans the bark so that dirt is not likely to be smeared on the cut surfaces of the seedlings.'

In apple trees the beneficial effect of vigorous inarches has been noticeable within three years, and thereafter the growth and health of trees previously stunted have greatly improved.

Pear trees in America which bore black-end fruits, because they were worked on *Pyrus pyrifolia*, were cured of this trouble by inarching with *P. communis* rootstocks. It was noticed[18] that the trees continued to bear black-end fruits until the inarches had completely taken over the work of the original roots and the *P. pyrifolia* had been entirely disconnected by severing below the junction of the inarches.

Apple trees severely damaged by winter frost have been saved for many years of cropping by the following procedure[33]. The frost-lifted rind was nailed down to prevent it gaping. Young trees, of hardy varieties tall enough to reach the branches, were planted by the S.W. side of each damaged tree. Shoots from the top of the young tree were then inarched into the branches of the older tree. It was found advisable to permit the development of a branch or two from the hardy variety to maintain a high degree of frost resistance in the new trunk. (*See also* page 237.)

CHAPTER IX

In Conclusion

Misuse of Grafting

The comparative ease with which two related plants may be intergrafted has encouraged nurserymen, more especially in continental Europe, to employ grafting in the multiplication of any plant which normally proves difficult by other means. This custom requires that the stock should come easily from seed, or root readily from cuttings or when layered. This has naturally led to the use of stocks whose chief attribute is their ready availability regardless of their suitability as rootstocks for satisfactory and lasting trees. Any gardener who has seen his lilacs (*Syringa* spp.) decline because they were worked on privet (*Ligustrum* spp.) or his choicest rhododendrons swamped by vigorous suckers from the *R. ponticum* rootstock, will look askance at any grafted shrub. It is clear that he is justified in preferring the non-grafted plant, and those nurserymen who make a habit of grafting all subjects which in the slightest degree are shy to root, will have only themselves to thank if all such grafting is brought into disrepute and their customers look elsewhere for the pure article. The use of unsuitable rootstocks should not loom so large that it brings the use of rootstocks in general into disrepute. The proper use of the rootstocks lies in the provision of a benevolent foundation for the scion and their interaction with particular scions has enabled horticulturists to gain excellent control of their material.

Stock/Scion Interaction

Extensive studies of the effect of rootstock upon scion[60], and, on a smaller scale, of scion upon rootstock[59,117,121,132], have shown that the rootstock, under given conditions, controls the size and habit of growth of the tree, time and degree of fruit-bud formation and setting, fruit colour, keeping qualities, and resistance to disease, and has been shown[49] to influence the rooting of cuttings taken from the scion. Rootstocks can be chosen for their resistance to adverse conditions of soil and climate and to suit selected modes of culture. This is not the place to go further into these matters; those interested should consult the relevant literature.

The grafted tree being composed of two parts, rootstock and scion, it is natural to expect that each part will affect the other. The effect of the rootstock has been well proved, but the effect of the scion upon the root has not been established to the same degree, partly, no doubt, because of the practical difficulty of observing the roots *in vitro*.

It has been shown[123] that the growth of a composite tree is mainly a compromise between the growth rates of rootstock and scion, provided the components are fully compatible. A dwarf scion on a vigorous rootstock makes a larger tree than does the vigorous scion on the dwarf, clearly indicating that the rootstock effect is greater than the scion effect (Plate 33).

The growth of a given scion achieves a degree of equilibrium with the root-system, but the proportion of shoot to root varies considerably from soil to soil, and on a sandy soil, lacking nutrients, twice as much root may be balanced against a unit of shoot as on a richer soil. It is clear that there is interaction not only between rootstock and scion but also between soil and climate, and in choosing rootstocks the tree-raiser must bear in mind the effects of all these factors. The use of intermediate stem pieces in the construction of trees greatly complicates the situation and it is not surprising that little definite knowledge has accrued relating to the interaction of rootstock, intermediate, and upper scion[118]. Where the unions with the intermediate exhibit no trace of incompatibility the effect of all three can have full play, but in these circumstances the intermediate piece is unlikely to have any very striking effect beyond, in certain combinations, circumventing incompatibility—of the breakage type—between rootstock and upper scion, though there have been indications that any such effect increases with the length of the intermediate. Thus, if the intermediate forms a large part

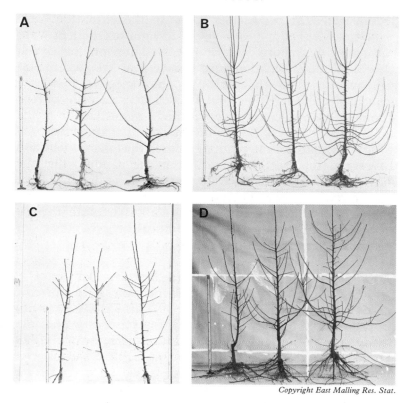

33. Rootstock/scion effect in apple

A. Dwarf M.9 worked on itself is dwarf.   B. Vigorous M.12 on itself is vigorous.   C. 12 on 9 is dwarfed.   D. 9 on 12 is invigorated. Thus the rootstock has the major influence.

of the whole tree, then it may be expected to contribute more to the general result. If the growth rate of the intermediate does not differ markedly from that of the upper scion, then it will not much affect the growth of the whole tree.

On the other hand, where the intermediate is not fully compatible with one of the other components, or forms structural complications at the graft junction, the intermediate may remain the limiting factor for growth of the tree. It seems feasible to suppose that this occurs when an extremely dwarf-growing intermediate is employed. Such an effect has been obtained in apple trees[122], and has been clearly demonstrated in pears (Plate 19). In the latter the combination of pear/pear/pear resulted in vigorous shoot growth, a coarse vertical root system and delayed blossom formation, whereas the combination pear/quince/pear resulted in less vigorous shoot growth, a fibrous spreading root system, and profuse blossoming. This inhibiting effect on growth, with its accompaniments, could be attributed to constriction at the union rather than to the free imposition of characteristics of the intermediate, or interaction between the intermediate and the rest of the tree. It should be borne in mind that the contributory effect of an intermediate, different in nature from the part above and below, will be the sum of a number of factors. These include not only its relative size and inherent properties, but also the physical and possibly other effects of the two additional graft unions.

Graft-Hybrids, Chimaeras, and 'Specific Influence'

Before concluding, it is necessary to mention, at least briefly, the subject of so-called graft-hybrids. This term leads one to suppose that hybridization can be accomplished by grafting. Such is not the case. In all probability no intermingling of characters has ever occurred. What has happened is that two subjects, similar in botanical characters, have joined together at their growing points. This may occur when a bud is formed right at the junction of stock and scion. Plants having such intermixed or overlaid tissues are termed chimaeras.

Winkler[129] has made a very full study of so-called graft-hybrids and was one of the first to produce these phenomena deliberately, though many ancient works suggest the possibility. For example, Fray Miquel Agustin wrote in the *Book of the Secrets of Agriculture*, printed in Madrid in 1695, 'You will make two bud shields, one from sour orange, the other from sweet orange, and from both you will form one . . . and they will make only one union which will make one bud to give the fruit half sweet and half sour.'

Winkler considered graft-hybrids of three classes: (1) those due to specific influence; (2) periclinal, sectorial, and mixed chimaeras; (3) cell-fusion graft-hybrids.

Specific influence, 'a permanent change in one component brought about by association with another, and such changed form transmissible by vegetative propagation', has never been proved to be other than temporary nutritive influence. In connection with this it is of some interest to recall that workers in nurseries have considered that growths from bud grafts which have lain dormant a long time sometimes temporarily bear a superficial resemblance to the rootstock in which they are inserted. Aspects of 'specific influence' have been studied in considerable detail by the Russian worker Michurin[90] who terms the agencies concerned 'mentors'.

Michurin and his followers claim to have modified the development of characters in young seedlings by grafting with well-established mature varieties to act as mentors to the young plants still in a 'plastic' state. The work of this group of Soviet horticulturists has received little support in scientific circles but, nevertheless, has intrigued many. Possibly of some interest in this connection is the more ready hybridization of wheat and rye plants that have been raised from transplanted embryos. Plants raised from wheat embryos on rye endosperm yield more hybrid kernels than do plants from wheat embryos on wheat endosperm[56].

Chimaeras have been found and made almost at will. Interesting examples are *Cytisus adamii* (or *Laburnum adamii*) which originated near Paris in 1825. It is said to have arisen after budding a *Laburnum vulgare* stock with *C. purpureus*. This bears flowers of yellow or purple and also a mixture of each. *Crataegomespilus dardari*, found near Metz in 1899, is a mixture of *Mespilus germanica* and *Crataegus monogyna*. There is no proof that cell-fusion hybrids, arising as a direct result of grafting, exist.

Grafting as an Aid to the Investigator

In addition to the extensive use of grafting in the rootstock studies of Hatton and other workers, grafting has been used in attempts to elucidate both physiological and pathological, fundamental problems. The amalgamation of rootstocks by grafting has served to clear up some problems of the source of rootstock effect.

All plant viruses are transmissible by grafting (Plate 34)[112] and this technique has proved invaluable in testing for the presence of disease, in the absence of symptoms, and also in

34. Virus transmission by grafting

A. A blackcurrant bush grafted with healthy scions has remained healthy.   B.
A similar bush grafted with one scion (on the left) infected with a virus disease
and showing typical symptoms, i.e. leaves with coarsely-toothed margins and
few veins. The disease has passed into the stock and the sucker shoot on the right
is now exhibiting similar symptoms.

proving the health of plants. For virus transmission by
grafting, 'organic union' is apparently vital, but it is also true
that a successful union, as understood by grafters, is in no way
essential for such a transmission, as is shown by the report of
virus transmission between diseased and healthy monocotyledonous plants (page 49), for example *Lilium auratum*.
Doubtless a close interlocking of callus tissue is often a
sufficient bridge.

Grafting is already increasingly employed in plant hormone and propagation studies, and might well be used more
widely by investigators in the elucidation of many other
problems.

Invention of Grafts
It appears that modes of grafting are continually rediscovered
and it is doubtful if any method can be entirely new. There

have been vain attempts to patent certain methods, but such presumption savours of the impertinent. By careful consideration of past achievement and appraisal of the results of biological research, itself based on a deep appreciation of the immutable laws of physics and chemistry, important developments in the use of grafting will be achieved. Plant propagators are already aware of the interplay of factors in obtaining maximum production; grafters also must appreciate the vital importance of the basic physical factors of gravity, pressure, orientation and spatial relations; and the chemical factors involved in diffusion, transport and influence of nutrients and messengers.

As remarked earlier, the methods of grafting are many, variations apparently unlimited, and it seems appropriate to repeat what Francis Drope of Oxford so fittingly wrote in 1672: 'These are the chief ways of grafting, some whereof are necessary, for a complete grafter to know, other some mere curiosities; but there are other variations which I purposely omit; supposing that from these, as from the chief heads, an ingenious lover of this art, will of his own industry, discover and improve them, to his greater pleasure, and content.'

References

1 ADDISON, G. and TAVARES, R. (1952): 'Hybridization and grafting in species of *Theobroma* which occur in Amazonia', *Evolution*, **6**, 380–6.

2 ARGLES, G. K. (1937): 'A review of the literature on stock/scion incompatibility in fruit trees, with particular reference to pome and stone fruits', *Imperial Bureau of Fruit Production, Technical Communication*, **9**.

3 BALTET, C. (1910): *The art of grafting and budding*, 6th ed., Crosby Lockwood, London.

4 BAYLEY BALFOUR, I. (1913): 'Problems of propagation', *Journal of the Royal Horticultural Society*, **38**, 447–60.

5 BAZAVLUK, V. Y. (1940): (The process of fusion and chimaera formation in potato grafts) (Russian) *Bulletin of the Academy of Sciences of the U.S.S.R., Biological Series*, **2**, 181–97.

6 BENNETT, H. D. (1926): 'An account of an experiment with pieceroot grafts in the apple and pear and of the influence on growth of the position of the top bud of the scion relative to the graft union', *Proceedings of the American Society for Horticultural Science*, **23**, 255–9.

7 BENNETT, M. (1972): 'Epidemiology—infection in the nursery', *Report of the East Malling Research Station for 1971*, **18**.

8 BITANCOURT, A. A. and FAWCETT, H. S. (1944): 'Statistical studies of distribution of psorosis-affected trees in citrus orchards', *Phytopathology*, **34**, 358–75.

9 BLUNDELL, J. B. (1976): 'Rose rootstock production' (*Proceedings*), *12th Refresher Course, Pershore, U.K.*, 14–17.

10 BRINGHURST, R. S. and VOTH, V. (1956): 'Strawberry virus transmission by grafting excised leaves', *Plant Disease Reporter*, **40**, 596–600.

11 BROWN, C. L. and SAX, C. (1962): 'The influence of pressure on the differentiation of secondary tissues', *American Journal of Botany*, **49**, 683–91.

12 BUCHHOLZ, J. T., DOAK, C. C. and BLAKESLEE, A. F. (1932): 'Control of gametophytic selection in *Datura* through shorten-

REFERENCES

ing and splicing of styles', *Bulletin of the Torrey Botanical Club*, **59**, 109–18.

13 BÜSGEN, M. (1929): *The structure and life of forest trees* (revised by M. Münch, translated by T. Thomson), Chapman & Hall, London.

14 CADMAN, C. H. (1940): 'Graft-blight of lilacs', *Gardeners' Chronicle*, **107**, 25.

15 CANNON, H. B. (1941): 'Studies in the variation of nursery fruit trees on vegetatively raised rootstocks', *Journal of Pomology*, **19**, 2–33.

16 CAVAZZA, D. (1923): *Viticoltura*. *Nuova enciclopedia agraria Italiana*, **5**, Unione Tipogratico—Editrice Torinese, Torino, 218–20.

17 CHANDLER, W. H. (1925): *Fruitgrowing*, Constable, London.

18 — (1942): *Deciduous orchards*, H. Kimpton, London.

19 CHANG, W. T. (1937): 'Studies in incompatibility between stock and scion with special reference to certain deciduous fruits', Ph.D. thesis, London University, pp. 151.

20 CURTIS, O. F. and BLAKE, J. H. (1936): 'The temperature of grafts as influenced by the type of wax and shading', *Proceedings of the 27th Annual Meeting, Northern Nut Growers' Association*, 41–4.

21 DARWIN, C. (1875): *The variation of animals and plants under domestication* (2nd Popular ed. used (1905), 2, 161), Murray, London.

22 DE STIGTER, H. C. M. (1956): 'Studies on the nature of the incompatibility in a cucurbitaceous graft', *Mededelingen Landbouwhogeschool Wageningen*, **56**, (8), 1–51.

23 — (1957): 'De kieming van rozenonderstammenzaad' (The germination of rose seeds for stocks), *Mededelingen Directeur van de Tuinbouw*, **20**, 356–62.

24 DIAS, C. E. A. (1928): 'Bud-grafting of rubber', *4th Quarterly Circular for 1928, Rubber Research Scheme (Ceylon)*, 6–7.

25 DROPE, F. (1672): *A short and sure guide in the practice of raising and ordering of fruit trees*, Oxford.

26 DURHAM, H. E. (1926): 'Stock or wedge grafting', *Gardeners' Chronicle*, **79**, 176–7.

27 EAMES, H. J. and COX, L. G. (1945): 'A remarkable tree-fall and an unusual type of graft-union failure', *American Journal of Botany*, **32**, 331–5.

28 EDGAR, A. T. (1958): *Manual of rubber planting (Malaya)*, Incorporated Society of Planters, Kuala Lumpur.

29 EMDEN, J. H. VAN. (1940): 'Een niewe entmethode voor thee' (A

new method of grafting tea), *Archief voor de Theecultuur in Nederlandsch-Indië*, **14**, 16–25.

30 EVANS, A. M. and DENWARD, T. (1955): 'Grafting and hybridization experiments in the genus *Trifolium*', *Nature* (U.K.), **175**, 687–8.

31 EVANS, H. (1953): 'Recent investigations on the propagation of cacao', *Report on Cacao Research, Imperial College of Tropical Agriculture for 1945–51*, 29–37.

32 FILEWICZ, W. (1938): 'Leczenie i wzmacnianie jabloni' (Bridge grafting and invigorating apple trees), *Roczniki Nauk Ogrodniczych*, **5**, 35–140.

33 — and MODLIBOWSKA, I. (1957): 'The hardiness of the south-western part of the tree as the major factor in its survival during test winters', *Report of the 14th International Horticultural Congress, Scheveningen, 1955*, 879–86.

34 FLEISCHHAUER, O. (1957): 'Ein neuartiger Okulationsschnellverschluss' (A new quick method for binding budded plants), *Deutsche Baumschule*, **9**, 23–5.

35 FLOOR, J. (1957): 'Moisture as a factor in the rooting of cuttings', *Report of the 14th International Horticultural Congress, Scheveningen, 1955*, 1140–8.

36 — (1971): 'Nieuwe methoden van vermeerdering' (New methods of propagation), *Bedrijfsontwikkeling, Editie Tuinbouw*, **2** (2), 43–8, 36.

37 GARNER, R. J. (1935): 'Studies in nursery technique. Shield budding—the removal of the wood', *Annual Report of the East Malling Research Station for 1934*, 123–6.

38 — (1943): 'Branch grafting for production of fruit tree pollinators', *Agriculture* (U.K.), **50**, 89–92.

39 — (1944): 'Double-working and bridging incompatible combinations of pear and quince', *Annual Report of the East Malling Research Station for 1943*, 80–5.

40 — (1944): 'Propagation by cuttings and layers. Recent work and its application, with special reference to pome and stone fruits', *Imperial Bureau of Horticulture and Plantation Crops, Technical Communication*, **14**.

41 — (1950): 'Studies in framework grafting of mature fruit trees. V. Fifteen years' comparative performance of frameworked and topworked apple trees', *Annual Report of the East Malling Research Station for 1949*, 71–4.

42 — (1951): 'The grafting of very young apple seedlings', *Annual Report of the East Malling Research Station for 1950*, 71–5.

43 — (1953): 'Double-working pears at budding time', *Annual Report of the East Malling Research Station for 1952*, 174–5.

44 — (1972): 'Dwarfing rootstocks and interstems for sweet cherries', *Proceedings of the 2nd Cherry Congress, Verona, 1972*, 101–11.

45 — (1976): 'Staking in relation to growth and form', *Combined Proceedings, International Plant Propagators' Society*, **25**, 210–14.

46 — and BEAKBANE, A. B. (1968): 'A note on the grafting and anatomy of black pepper', *Experimental Agriculture*, **4**, 187–92.

47 — and HAMMOND, D. H. (1938): 'Studies in incompatibility of stock and scion. II. The relation between time of budding and stock/scion compatibility', *Annual Report of the East Malling Research Station for 1937*, 154–7.

48 — and — (1939): 'Studies in nursery technique. Shield budding. Treatment of inserted buds with petroleum jelly', *Annual Report of the East Malling Research Station for 1938*, 115–17.

49 — and HATCHER, E. S. J. (1957): 'Rootstock effect on the regeneration of apple cuttings', *Annual Report of the East Malling Research Station for 1956*, 60–2.

50 — and — (1963): 'Regeneration in relation to vegetative vigour and flowering', *(Proceedings of the) 16th International Horticultural Congress, Brussels, 1962*, 105–11.

51 — and NICOLL, C. P. (1961): 'The compatibility and growth of pear scions on apple rootstocks', *Report of the East Malling Research Station for 1960*, 54–6.

52 — and WALKER, W. F. (1938). 'The frameworking of fruit trees', *Imperial Bureau of Horticulture and Plantation Crops, Occasional Paper*, **5**.

53 GLENN, E. M. (1966): 'Incompatibility in the walnut', *Report of the East Malling Research Station for 1965*, 102.

54 GRAHAM, B. F. and BORMANN, F. H. (1966): 'Natural root grafts', *Botanical Review*, **32**, 255–92.

55 GUENGERICH, H. W. and MILLIKAN, D. F. (1965): 'Root grafting, a potential source of error in apple indexing', *Plant Disease Reporter*, **49**, 39–41.

56 HALL, O. L. (1954): 'Hybridization of wheat and rye after embryo transplantation', *Hereditas*, **40**, 453–8.

57 HANCOCK, W. G. (1940): 'Grafting male pawpaw trees', *Queensland Agricultural Journal*, **54**, 377–9.

58 HARRIS, R. V. and KING, M. E. (1942): 'Studies in strawberry virus diseases. V. The use of *Fragaria vesca* L. as an indicator of

REFERENCES

Yellow-edge and Crinkle', *Journal of Pomology*, **19**, 227–42.

59 HATTON, R. G. (1930): 'The relationship between scion and rootstock with special reference to the tree fruits', *Journal of the Royal Horticultural Society*, **55**, 169–211.

60 — (1935): 'Apple rootstock studies. Effect of layered stocks upon the vigour and cropping of certain scions', *Journal of Pomology*, **13**, 293–350.

61 — (1936): 'Plum rootstock studies: their effect on the vigour and cropping of the scion variety', *Journal of Pomology*, **14**, 293–350.

62 — and BAGENAL, N. B. (1934): 'Scion-rooting at East Malling Research Station', *Annual Report of the East Malling Research Station for 1933*, 243–5.

63 HEARMAN, J., BEAKBANE, A. B., HATTON, R. G. and ROACH, W. A. (1936): 'The reinvigoration of apple trees by the inarching of vigorous rootstocks', *Journal of Pomology*, **14**, 376–90.

64 HERRERO, J. (1951): 'Studies of compatible and incompatible graft combinations with special reference to hardy fruit trees', *Journal of Horticultural Science*, **26**, 186–237.

65 HILKENBÄUMER, F. (1942): 'Die gegenseitige Beeinflussung von Unterlage und Edelreis bei den Hauptobstarten im Jugendstadium unter Berücksichtigung verschiedener Standortsverhältnisse' (The reciprocal influence of stock and scion in the chief fruit types in their early stages with special reference to locality), *Kühn-Archiv*, **58**, pp. 261.

66 HODGSON, R. W. (1923): 'Saving the gophered citrus tree', *Circular, California Agricultural Experiment Station*, **273**.

67 HOWARD, B. H. (1974): 'Chip budding', *Report of the East Malling Research Station for 1973*, 195–7.

68 —, SKENE, D. S. and COLES, J. S. (1974): 'The effects of different grafting methods on the development of one-year-old nursery apple trees', *Journal of Horticultural Science*, **49**, 287–95.

69 HUROV, H. R. and CHONG, Y. F. (1960): 'Illustrated description of the green bud strip budding of 2–8 month old rubber seedlings', *Pamphlet, Department of Agriculture, North Borneo*, **9**.

70 INGRAM, C. (1942): 'A new method of graftage', *Gardeners' Chronicle*, **112**, 186.

71 JAFFE, A. (1970): 'Chip budding guava cultivars', *Plant Propagator*, **16** (2), 6.

72 JAYNES, R. A. (1964): 'Grafting chestnuts without stock plants', *55th Annual Report of the Northern Nut Growers' Association*, 16–20.

REFERENCES

73 JONES, O. P., HOPGOOD, M. E. and O'FARRELL, D. (1977):
'Propagation *in vitro* of M.26 apple rootstocks', *Journal of Horticultural Science*, **52**, 235–8.

74 KEANE, F. W. L. and MAY, J. (1963): 'Natural root grafting in cherry, and spread of cherry "twisted leaf virus"', *Canadian Plant Disease Survey*, **43**, 54–60.

75 KERR, W. L. (1936): 'A simple method of obtaining fruit trees on their own roots', *Proceedings of the American Society for Horticultural Science*, **33**, 355–7.

76 KNIGHT, T. A. (1812): 'A new and expeditious mode of budding', *Transactions of the Horticultural Society, London*, **1**, 194–6.

77 KUMAR, D. and WARING, P. F. (1973): 'Studies in tuberization in *Solanum andigena*. 1. Evidence for the existence and movement of a specific tuberization stimulus', *New Phytologist*, **72**, 283–7.

78 KUNTZ, J. E. and RIKER, A. J. (1956): 'Oak wilt', *Bulletin, Wisconsin University Agricultural Experiment Station*, **519**.

79 LANGFORD, M. H. *et al.* (1954): '*Hevea* diseases of the Western Hemisphere', *Plant Disease Reporter*, **225** (Supplement), 37–41.

80 LA RUE, C. D. and MUZIK, T. J. (1954): 'Growth, regeneration, and precocious rooting in *Rhizophora mangle*', *Michigan Academy of Sciences, Arts, Letters*, **39**, 9–29.

81 LESLIE, W. R. (1938): *Dominion Experimental Station, Morden, Manitoba, Results of Experiments 1931–7*.

82 LOEB, J. (1924): *Regeneration from a physico-chemical viewpoint*, McGraw Hill, New York.

83 LOWMAN, M. S. and KELLY, J. W. (1946): 'The presence of mydriatic alkaloids in tomato fruit from scions grown on *Datura stramonium* rootstock', *Proceedings of the American Society for Horticultural Science*, **48**, 249–59.

84 LYNCH, S. T. and NELSON, R. (1950): 'Mango budding', *Proceedings of the Florida State Horticultural Society for 1949*, **62**, 207–9.

85 MARSTON, M. E. (1953): 'The history of plant propagation in England to 1850' (thesis, Nottingham University).

86 MAWE, T. and ABERCROMBIE, J. (1778): '*The universal gardener and botanist or, a general dictionary of gardening and botany*', London.

87 MENDEL, K. (1936): 'The anatomy and histology of the bud-union in citrus', *Palestine Journal of Botany and Horticultural Science*, **1** (2), 13–46.

88 MERRILL, S., JR. (1944): 'The budding of tung (*Aleurites fordii*

Hemsl.)', *Proceedings of the American Society for Horticultural Science*, **44**, 227–35.

89 METCALFE, C. R. and CHALK, L. (1950): *Anatomy of the dicotyledons*, Clarendon Press, Oxford.

90 MICHURIN, I. V. (1939): (The use of 'mentors' in raising hybrid seedlings and examples of definite changes induced in fruit tree varieties by various external factors) (Russian) in *Michurin's selected works*, Voronezh Region Publishers, Voronezh, 202–23.

91 MILLER, P. W. *et al.* (1957): 'Blackline of Persian walnut', *48th Annual Report of the Northern Nut Growers' Association*, 32–3.

92 MILLNER, M. E. (1932): 'Natural grafting in *Hedera helix*', *New Phytologist*, **31**, 2–25.

93 MOORE, J. C. (1963): 'Propagation of chestnuts and cammellia by nurse seed grafts', *Combined Proceedings of the International Plant Propagators' Society*, **13**, 141–3.

94 MOSSE, B. (1962): *Graft incompatibility in fruit trees*, Commonwealth Agricultural Bureaux, Farnham Royal.

95 — and GARNER, R. J. (1954): 'Growth and structural changes induced in plum by an additional scion', *Journal of Horticultural Science*, **29**, 12–20.

96 MUZIK, T. J. and LA RUE, C. D. (1952): 'The grafting of large monocotyledonous plants', *Science* (U.S.A.), **116**, 589–91.

97 NATTRASS, R. M. (1944): 'The transmission of the virus of the "woodiness" disease of passion fruit *(Passiflora edulis)* by single-leaf grafts', *Annals of Applied Biology*, **31**, 310–11.

98 NEEL, P. L. and HARRIS, R. W. (1971): 'Motion-induced inhibition of elongation and induction of dormancy in *Liquidambar*', *Science* (U.S.A.), **173**, 58–9.

99 NICOLIN, P. (1953): 'Das "Nicolieren", eine neue Veredlungsmethode' ('Nicolining', a new budding method), *Deutsche Baumschule*, **5**, 186–7.

100 PEROLD, A. I. (1927): *A treatise on viticulture*, Macmillan, London.

101 PESCOTT, E. E. (1934): 'Changing the variety of fruit trees—some modern methods', *Journal of the Department of Agriculture of Victoria*, **32**, 517–22, 535.

102 PLOTNIKOV, I. G. (1940): (Vegetative hybridization of cereals and its importance in breeding work) (Russian) Sotsialisticheskoe Zernovoe Khozyaistvo, **6**, 69–88.

103 POSNETTE, A. F. (1953): 'Virus diseases and the propagation of fruit trees', *Annual Report of the East Malling Research Station for 1952*, 179–81.

REFERENCES

104 REYNAUD, M. (1944): 'Le greffage des oléastres en Algérie' (Grafting wild olives in Algeria), *Bulletin, Direction de l'Agriculture d'Algerie*, **98**.

105 REYNA, E. H. (1966): 'Un nuero método de injertación en café' (A new method of grafting coffee), *Boletín Técnico, Dirección General de Investigacion y Control Agropecuario, Ministerio de Agricultura, Guatemala*, **21**.

106 ROACH, W. A. (1930): 'Increased scion vigour induced by certain foreign rootstocks', *Annals of Botany*, **44**, 859–64.

107 SANSOME, F. W. (1957): private communication—University College, Ibadan.

108 SHAMEL, A. D., POMEROY, C. S. and CARYL, R. E. (1925): 'Citrus fruit-growing in the South-West', *Farmers' Bulletin, U.S. Department of Agriculture*, No. 1,447.

109 SHARPLES, A. and GUNNERY, H. (1933): 'Callus formation in *Hibiscus rosa-sinensis* L. and *Hevea brasiliensis* Müll. Arg.', *Annals of Botany*, **47**, 827–40.

110 SHEFFIELD, F. M. L. (1952): 'Studies on the clove tree. IV. Natural grafting and its bearing on sudden-death disease', *Annals of Applied Biology*, **39**, 103–10.

111 SITTON, B. G. (1940): 'Pecan grafting methods and waxes', *Circular, U.S. Department of Agriculture*, **545**.

112 SMITH, K. M. (1933): '*Recent advances in the study of plant viruses*', Churchill, London.

113 SNYDER, E. and HARMON, F. N. (1939): 'The bud graft method of propagating *vinifera* grape varieties on rootstocks', *Leaflet, U.S. Department of Agriculture*, **173**.

114 SNYDER, W. E. and HESS, C. E. (1957): 'An evaluation of the mist technique for the rooting of cuttings as used experimentally and commercially in America', *Report of the 14th International Horticultural Congress, Scheveningen, 1955*, 1125–32.

115 STARKER, T. J. (1942): 'Will girdling always kill Douglas fir?', *Journal of Forestry*, **40**, 658–9.

116 STOUTEMYER, V. T. (1937): 'Regeneration in various types of apple wood', *Research Bulletin, Iona Agricultural Experiment Station*, **220**.

117 SWARBRICK, T. and ROBERTS, R. H. (1927): 'The relation of scion variety to character of root growth in apple trees', *Research Bulletin, Wisconsin Agricultural Experiment Station*, **78**.

118 —, BLAIR, D. and SINGH, S. (1946): 'Studies in the physiology of rootstock and scion relationships', *Journal of Pomology*, **22**, 51–61.

119 THOUIN, A. (1821): *Monographie des greffes, ou description technique*, Paris.

120 TRUNK, C. (1932): 'A serious trouble of walnuts', *18th Annual Report of the Western Nut Growers' Association, Oregon*, 163–8.

121 VYVYAN, M. C. (1930): 'The effect of scion on root. III. Comparison of stem- and root-worked trees', *Journal of Pomology*, **8**, 259–82.

122 — (1938): 'The relative influence of rootstock and of an intermediate piece of stock in some double-grafted apple trees', *Journal of Pomology*, **16**, 251–73.

123 — (1955): 'Interrelation of scion and rootstock in fruit trees', *Annals of Applied Biology*, **19**, 401–23.

124 WALLACE, J. M. (1947): 'The use of leaf tissue in graft transmission of psorosis virus', *Phytopathology*, **37**, 149–52.

125 WARD, J. M. (1929): 'Budding, grafting, and inarching of fruit trees', *Bulletin, Department of Agriculture, Victoria*, **63**.

126 WEBBER, G. G. and MACDANIELS, L. H. (1969): *North American nut trees*, Northern Nut Growers' Association, Knoxville, Tennessee, 18–38.

127 WETMORE, R. H. and SOROKIN, S. (1955): 'On the differentiation of xylem', *Journal of the Arnold Arboretum*, **36**, 305–17.

128 WILLIAMS, W. (1959): 'Pear-apple hybrids', *Report of the John Innes Horticultural Institution for 1958*, **8**.

129 WINKLER, H. (1907): 'Über Pfropfbastarde und pflanzliche chimären' (On graft-hybrids and plant chimaeras), *Bericht der Deutschen Botanischen Gesellschaft*, **25**, 568–76.

130 WOODWARD, J. S., ROMBERG, L. D. and WILLMANN, F. L. (1929): 'Pecan growing in Texas', *Bulletin, Texas Department of Agriculture*, **95**.

131 WORMALD, H. (1943): 'Papery bark canker of fruit trees in relation to silver-leaf disease', *Journal of Pomology*, **20**, 144–6.

132 — and GRUBB, N. H. (1924): 'The crown gall disease of nursery stocks. I. Field observations on apple stocks', *Annals of Applied Biology*, **11**, 278–91.

133 YASUDA, S. (1931): 'An experiment to graft the style upon the ovary in *Petunia violacea*', *Proceedings of the Imperial Academy of Japan*, **7**, 72–5.

134 YOUNG, H. E. (1954): 'Crown budding for oidium resistance', *Advisory Circular, Rubber Research Institute of Ceylon*, **32**.

135 ZAAR, E. I. (1953): 'Grafting of *Quercus robur*, using the plumule of a germinating seed as scion', *Botanicheskii Zhurnal*, **38**, 440–1.

APPENDIX I

PEARS ON QUINCE ROOTSTOCKS

List of varieties said to be more reliable when double-worked (Beurré Hardy or Fertility suggested for intermediates).
Bartlett.
Belle Julie.
Beurré Bosc.
Beurré Clairgeau.
Beurré St. Nicholas (Duchesse d'Orléans).
Bristol Cross.
Citron des Carmes (Madeleine).
Clapp's Favourite.
Comte de Lamy.
Dr. Jules Guyot.
Dr. Hogg.
Doyenné d'Eté.
Early Market.
Fondante de Charneau (Legipont).
Fondante de Thirriott.
Forelle (Trout Pear).
Jargonelle.
Marguerite Marillat.
Marie Louise.
Merton Pride.
Monarch.
Nec Plus Meuris.
Packham's Triumph.
Reine des Poires.
Sandar William Crouzee.
Souvenir de Congrès.
Williams' Bon Chrétien (Bartlett).*

* Selections of Williams' compatible with quince are now available and are replacing the incompatible clones in Britain and elsewhere.

PLUM ROOTSTOCKS AND THEIR USES

Some Suggested Combinations of Rootstock and Scion which appear Suitable and Compatible, and some to be Avoided

Scion Varieties	Rootstocks						
	Myrobalan B	Brompton	Marianna	St. Julien A	Common Plum	Common Mussel	Pershore
Belle de Louvain	L	L	L			L	
Bryanstone Gage	L	L					
Bush (Kentish)	L				S		
Cambridge Gage	L		L	S			
Cherry Plum	X			S			S
Comte d'Althann's Gage	L	L	X		S	S	
Czar	L	L	X		X	S	
Damsons	L		X		X		
Denniston's Superb Gage	L				S	S	
Early Laxton	L		L	S	S	S	
Giant Prune	L		L		S	S	
Jefferson	L						
Laxton's Gage	X	L			S		
Marjorie's Seedling	L	L	X	S	X	S	
Monarch	L	L			S		S
Ontario	X	L					
Oullins Golden Gage	X	L	X	S	S	S	S
Pershore (Yellow Egg)	L	L	L		S	S	S
Pond's Seedling	L	L				S	
President	L	L	X		X	S	
Purple Pershore	L	L	L		S		
Rivers's Early Prolific	L	L	L		S	S	
Severn Cross	L	L			S		
Thames Cross	L	L			S		
Victoria	L	L		S	S	S	S
Warwickshire Drooper	L	L			S	S	
Peaches and Nectarines	X	L		S	X	L	
Apricots	L	L		S		L	

L = Combinations for large to medium trees.   S = Combinations for medium to small trees.   X = Combinations to avoid.

APPENDIX III

PLUM ROOTSTOCK/SCION COMBINATIONS TO AVOID

Myrobalan B is unreliable for Comte d'Althann's Gage, Marjorie's Seedling, Oullins Golden Gage and peaches.

Marianna is unreliable for Czar, Denniston's Superb Gage, Marjorie's Seedling, Oullins Golden Gage, President and damsons.

Common Plum is unreliable for Czar, Marjorie's Seedling, President, damsons and peaches.

Brussels should not be used for plums or peaches.

APPENDIX IV

REPLANT DISEASES AND THEIR CONTROL

The replanting of once-used nursery land with the same species or genera, even after intervals of many years, sometimes proves quite unremunerative owing to the effects of specific replant diseases. For further enlightenment see the extensive review by Dr B. M. Savory (1966): *Specific replant diseases*, published by Commonwealth Agricultural Bureaux, Farnham Royal, Bucks, England.

These replant diseases, whose cause is not fully known, damage the roots of newly-planted trees, rootstocks and cuttings of many species but most of all citrus, apple, cherry and peach and, to a lesser extent pear, plum (*P. domestica*) and rose (*R. canina*). As a result, growth in the first year after replanting may be so poor that the rootstocks may be unfit to work. Apples are affected not only on land that has carried apples but also where pears or quinces have been grown. Similarly cherries also suffer on both cherry and plum land. The effects on apples are rarely severe on acid soils, where the soil pH is 5·9 or less, but where the reaction is above 6·0 some avoiding action or amelioration will be necessary. Completely fresh land may well carry a single cycle of crops, provided the nutritive requirements of the crops are met, but thereafter it should be avoided for the same or closely allied species.

Where other considerations enforce the re-employment of old sites for susceptible crops consideration should be given to the use of partial soil sterilization by fumigation. To date, the only really successful soil fumigant to use for this purpose is chloropicrin, as little as 280 litres/hectare (25 gals/acre) proving satisfactory on nursery soils (see B. M. Savory (1967): 'Specific replant diseases of apple and cherry', *Annual Report of the East Malling Research Station for 1966*). Chloropicrin is a very dangerous chemical but it may be quite safely and legally used in the U.K. as a soil fumigant, provided it is applied in accordance with a very strict Code of Practice. The regulations covering its use deal with all aspects of application and the sealing of the fumigated soil with polyethylene sheeting. Chloropicrin fumigation must only be carried out by fully-qualified and trained personnel. If fumigation is done in early autumn (early September in the Northern Hemisphere) the soil will be free of noxious fumes and ready for planting in early winter, some two months after treatment. Properly carried out, fumigation with chloropicrin restores the productivity of the site; it may in some areas lead to even better growth than would have occurred in fresh ground.

In the case of cherries, specific replant disease has been found to be closely associated with the fungus *Thielaviopsis basicola*. This fungus tends to build up in the soil quite rapidly wherever cherry rootstocks are grown, and it is responsible for the rapid falling-off in production that so often occurs in cherry F.12/1 layer beds. Work in Holland and at East Malling has shown that a good control of *Thielaviopsis basicola* can be obtained by giving the F.12/1 cherry layer beds an annual drench, applied in early May, with the fungicide Benomyl.

APPENDIX V

Weed Control in Nurseries by Herbicides

As the growth of a young grafted shrub or tree in the nursery is decided largely by the growth of the rootstock before grafting, weed competition, especially severe in closely-planted rows, cannot be tolerated. Soil surface cultivations among young plants are too often accompanied by widespread damage from scraping and bruising; this damage is often progressive and leads to the complete loss of otherwise saleable trees. Such loss is avoided by the use of herbicides.

The nursery site should be free of perennial weeds. If not, weed control will prove difficult, if not hazardous. The efficient use of herbicides in tree and shrub nurseries is complicated by the wide number of species involved, some of which may be damaged by particular chemicals. Information should be sought from the official advisory services and from the suppliers. Addresses of the latter appear in the classified telephone directories and the trade press. The 'label' recommendations of the suppliers should be closely followed. If in doubt concerning particular materials, the suppliers should be consulted direct.

Nurseries infested with perennial weeds such as couch grass may be treated with propyzamide (Kerb 50W), but only when the plants have already been established in the open for at least one year. Particulars are obtainable from Pan Britannica Industries, Britannica House, Waltham Cross, Herts, England.

Two herbicides already proved generally useful in tree and shrub nurseries are simazine and paraquat.

Simazine is a residual soil-acting herbicide which may be used on leafless dormant plants, or non-rooted leafless cuttings, as an overall spray provided the recommended rates of application are not exceeded. The effects of simazine are influenced by soil type and rainfall, plants grown on sandy soils or under heavy rainfall being more liable to injury than those grown on clay or peat soils. Because

306

some species are less tolerant than others, care must be taken to avoid overdosing and it is essential to apply the material evenly with a suitable applicator. Simazine applied to weed-free moist soil at a rate of 1–2 kg of active ingredient per hectare (1–2 lb. per acre) will keep the nursery weed-free for several months. The higher rate should be used on the heavier soils.

Paraquat is a contact herbicide which must be directed on to the foliage to be killed and kept off the stems and foliage of the nursery material. Special equipment is available for applying paraquat between rows. Rates of application may vary considerably both in concentration and by volume for a given area but in general 1 kg of active ingredient per 500 litres of water sprayed on a hectare of weed-covered land (1 lb. per 100 gallons per acre) is adequate. Following application of simazine, it is normally only necessary to deal with isolated patches of weed by means of paraquat.

Index